2ND EDITION

THE COMPLETE BOOK OF
DUCATI MOTORCYCLES

EVERY MODEL SINCE 1946

Ian Falloon

Quarto

© 2022 Quarto Publishing Group USA Inc.
Text © 2016, 2022 Ian Falloon

Second edition published in 2022
First Published in 2016 by Motorbooks, an imprint of The Quarto Group,
100 Cummings Center, Suite 265-D, Beverly, MA 01915, USA.
T (978) 282-9590 F (978) 283-2742 Quarto.com

All rights reserved. No part of this book may be reproduced in any form without written permission of the copyright owners. All images in this book have been reproduced with the knowledge and prior consent of the artists concerned, and no responsibility is accepted by producer, publisher, or printer for any infringement of copyright or otherwise, arising from the contents of this publication. Every effort has been made to ensure that credits accurately comply with information supplied. We apologize for any inaccuracies that may have occurred and will resolve inaccurate or missing information in a subsequent reprinting of the book.

Motorbooks titles are also available at discount for retail, wholesale, promotional, and bulk purchase. For details, contact the Special Sales Manager by email at specialsales@quarto.com or by mail at The Quarto Group, Attn: Special Sales Manager, 100 Cummings Center, Suite 265-D, Beverly, MA 01915, USA.

26 25   6 7

ISBN: 978-0-7603-7373-6

Digital edition published in 2022
eISBN: 978-0-7603-7374-3

Formerly found under the following Library of Congress Cataloging-in-Publication Data

Names: Falloon, Ian, author.
Title: The complete book of Ducati motorcycles : every model since 1946 / by Ian Falloon.
Description: Minneapolis, Minnesota : Motorbooks, 2016. | Includes index.
Identifiers: LCCN 2015046240 | ISBN 9780760350225 (hc w/jacket)Subjects: LCSH: Ducati motorcycle--Pictorial works. | Ducati motorcycle--History.
Classification: LCC TL448.D8 F34926 2016 | DDC 629.227/5--dc23LC record available at http://lccn.loc.gov/2015046240

Cover Design: traffic
Interior Design: Simon Larkin
Cover Images: Front cover, Ducati; back cover top and middle, Ian Falloon; back cover bottom, Ducati; back flap, Ian Falloon; front flap, Ducati
Endpaper Images: Front endpaper, Ian Falloon; back endpaper, Ducati

Printed in China

THE COMPLETE BOOK OF DUCATI MOTORCYCLES

TABLE OF CONTENTS

4 ACKNOWLEDGMENTS

6 INTRODUCTION

8 **CHAPTER 1** After the War: Cucciolos and Overhead-Valve Singles, 1946–1954

22 **CHAPTER 2** Taglioni's Arrival: The Gran Sport and Desmodromics Overhead Camshaft and Overhead-Valve Singles, 1955–1960

58 **CHAPTER 3** Dubious Proliferation: Overhead Camshaft, Overhead-Valve Singles, and Two-Strokes, 1961–1970

100 **CHAPTER 4** Ducati's First Superbike: The 750, Singles, and Twins, 1971–1974

120 **CHAPTER 5** Economic Rationalization: The 860, 900, Parallel Twin, and Pantah, 1975–1984

148 **CHAPTER 6** The Cagiva Era: The Paso, 851, and 888, 1986–1993

172 **CHAPTER 7** Groundbreaking New Models: The 916 and Monster, 1994–2002

198 **CHAPTER 8** Confused Directions: The 999 and Sport Classic, 2003–2006

212 **CHAPTER 9** Grand Prix Success: MotoGP, Desmosedici, 1098, 2007–2011

232 **CHAPTER 10** A New Lease on Life: Panigale, MotoGP Resurrection, the Scrambler, and V4 2012–2022

278 INDEX

ACKNOWLEDGMENTS

Although I have written several books on Ducati Motorcycles, to comprehensively cover every model in one volume is a challenge. Information on many early Ducatis is difficult to obtain, requiring considerable time spent pouring through period motorcycle magazines and factory publications. Fortunately, Ducati Museo curator Livio Lodi came through with most of the production data from 1960, published here for the first time and significantly simplifying the categorization of the multitude of models built in the 1960s.

While I have a strong interest in the history of Ducati and all their motorcycles, my entry in the world of the *Ducatisti* coincided with the release of the 750GT. In the early seventies little was initially known about the new Ducati "L-Twin" but several significant magazine articles opened my eyes. The first was a 750GT test in the October 1972 *Cycle* magazine, followed over the next few years by the magazine's editors Cook Neilson and Phil Schilling continually extoling the Ducati's virtues on the road and track. Later I was fortunate to get to know Phil, and more recently Cook, and owe much of my enthusiasm for the marque to these two towering icons. I first approached Phil over twenty years ago for help with my first book *The Ducati Story*, and his mentoring and subsequent support was immeasurable. With Phil's passing, the Ducati world lost one of its most significant advocates.

As always, I have relied on the help of others, particularly with photos and information on some of the early models. My good friend Roy Kidney has always been forthcoming with high quality photos, and others I would like to thank are David James at Ducati, Alessandro Altinier, Phil Andersen, Rob Barker, Jerry Dean, Angus Dykman, Elvis Centofanti, Peter Hageman, Wolfgang Hobisch, Nico Georgeoglou, John Goldman, Don Kotchoff, Eric Kurtev, Fabio Lorenzet, Tim O'Mahoney, Ninni Pisciotta, Claudio Scalise, Allen Tannenbaum, Pierre Terblanche and Jon White.

At Motorbooks, my thanks go to Publisher Zack Miller who was the force behind this second edition. At home the support of my family, Miriam, Ben and Tim was undiminished as always.

—Ian Falloon, October 2021

INTRODUCTION

In 1922, when nineteen-year-old physics student Adriano Cavalieri Ducati was conducting his first experiments with radio, the production of motorcycles could not have been further from his mind. By 1924, from his house in Bologna, he had established radio contact with the United States, a considerable feat at the time, and on July 4, 1926, Adriano, with his two brothers Bruno and Marcello, set up a company, *Società Scientifica Radio Brevetti Ducati*, or SSR, building condensers and other radiographic components. While Adriano was a brilliant technician, Bruno assumed the role of manager and administrator, and Marcello that of designer. At this time Bologna was a center of radio development, with Guglielmo Marconi, the inventor of the radio, one of Bologna's most famous sons. Under the fascist dictator Mussolini's policy of modernization, the Ducati brothers' business expanded considerably, and by 1935 they had acquired a new site at Borgo Panigale, an industrial district on the outskirts of Bologna.

On this site, the Ducati brothers built a huge modern factory to manufacture radios and electronic components, the company's prosperity undoubtedly aided by Mussolini's interest in the widespread availability of radios to further his propaganda. By 1940, Ducati employed 11,000 workers and was the second largest company in Italy, with a large research department allowing diversification to include optical and mechanical divisions alongside the electronic. Ducati started to produce cameras and lenses, as well as cash registers and electric razors. The Raselet was the first electric razor produced in Italy, and because Italy was largely untouched by the war during the early years, the company continued to grow until the Italian government sought an armistice with the Allies on September 8, 1943.

That was when things started to go wrong for the Ducati brothers. On the day after the armistice, German soldiers commandeered the factory at Borgo Panigale, eventually transferring much of the plant and machinery to Germany. Further disaster occurred on October 12, 1944, when Allied bombing almost completely destroyed the factory. After the war, the Ducati brothers were briefly involved with Allied naval intelligence, but attempts to rebuild the company were stricken with financial problems caused by the expense of reconstruction and the general difficulties of the immediate postwar period. On December 1, 1947, the *Società Scientifica Radio Brevetti Ducati* went bankrupt, but the company was considered by the IRI (*Istituto di Ricostruzione Industriale*) too important to allow it to collapse completely.

A government and Vatican financial consortium came to the company's rescue, and in 1948 control of Ducati passed from the brothers to that of the FIM (*Fondo Industrie Meccaniche*) and IMI (*L'Istituto Mobiliare Italiano*). Now under government control, a mixture of good and bad political decisions would determine Ducati's history over the next forty years. In the meantime, the Borgo Panigale plant had been rebuilt, and Ducati entered into an alliance with SIATA (*Società Italiana Applicazioni Techiche Auto-Aviatore*), a Turin technical development company known for developing racing auto engines, to produce a clip-on motorcycle engine, the Cucciolo or Puppy. The factory in Borgo Panigale began producing Cucciolo motors, condensers, cameras, and cinema projectors, while Milan became the center for Ducati radios and electronics. The high-quality 18x24mm format rangefinder cameras, with their range of screw-mount lenses, was hailed as a miniature Leica, but it was the Cucciolo that would secure Ducati's future. From these humble beginnings grew one of Italy's greatest motorcycle makes.

CHAPTER 1
AFTER THE WAR
CUCCIOLOS AND OVERHEAD-VALVE SINGLES, 1946–1954

Beginning in June 1946, Ducati also built the Cucciolo T1.

The Cucciolo was initially produced by SIATA in Turin as a motor to attach to a bicycle.

After the armistice in 1943, Turinese lawyer and writer Aldo Farinelli designed a prototype auxiliary engine that could be easily mounted on a bicycle. Against government directives, he had a prototype running by 1944 and nicknamed it the Cucciolo because of the high-pitched, yapping exhaust. The Cucciolo seemed an unremarkable design, but the immediate postwar period was a very opportune time to produce such a motor. It was easily installed in a bicycle frame, providing cheap and economical transportation, and because it was a four-stroke, it was extremely economical. Fuel was a scarce commodity in postwar Italy, and Farinelli's little engine had the added bonus of good torque and two-speed gearing, allowing the bicycle to climb modest grades.

CUCCIOLO T1

Farinelli's design was adopted by SIATA (*Società Italiana Applicazioni Techiche Auto-Aviatore*) in Turin, and in July 1945, barely a month after the end of World War II, the company announced plans to produce the Cucciolo. As the first new automotive design to appear in postwar Europe, demand soon outstripped supply and SIATA looked for a partner with more production capability. It turned to Ducati, and in June 1946 Ducati also began to supply Cucciolo engines. At this stage, Ducati was still called SSR Ducati (*Società Scientifica Radio Brevetti Ducati*), the distinctive SSR logo still adorning its products. These early Ducati-produced Cucciolo T1 engines were identical to the SIATA examples, with diagonal cylinder finning and the 20-degree inclined forward alloy cylinder cast integrally with the narrow vertically split aluminum crankcase. The steel con rod ran in roller bearings, with two ball bearings supporting the crankshaft.

The cylinder head design was unusual, with both the exhaust and intake facing rearward. The two overhead valves were parallel and operated by twin pull rods, rather than the usual pushrods, behind the cylinder and driven by a single cam in the crankcase. The rockers and valve springs were exposed, and although only lubricated by gravity drip feed, engine life was exceptional, and the Cucciolo soon earned a reputation for outstanding reliability.

A gear primary drive drove a wet multiplate metallic clutch, and a preselector allowed the pedals to shift gears. The pedals in a vertical position selected neutral, with the right pedal forward selecting first and the left pedal forward selecting second gear. A lever on the left handlebar, with a hand throttle on the right handlebar, operated the clutch.

The T1 engine was designed to attach to a bicycle frame, using the bicycle chain for a secondary drive, and included a 2-liter gas tank that fitted above the rear wheel. Starting was by the usual moped method of pedaling while the engine was in gear. With claims of 275 miles per gallon, the Cucciolo was so successful that production levels continued to increase during 1947, to more than 240 a day, although at this stage the Cucciolo was still only available in Italy. By 1947, with more than 25,000 sold, Ducati assumed control of the entire manufacture and distribution, establishing an export department in Milan. To celebrate the launch, Ducati commissioned Maestro Olivero to write a song, this becoming a hit in Italy, and so confident was Ducati that engineer Aldo Loria took one to New York in 1947. The Cucciolo didn't end up conquering America, but it was widely distributed throughout the world and was the most successful of all the postwar micromotors.

CUCCIOLO T1 1946–1947

Engine	Single-cylinder four-stroke, air-cooled
Bore (mm)	39
Stroke (mm)	40
Displacement (cc)	48
Compression ratio	6.25:1
Valve type	Overhead valve, pullrod
Carburetion	Weber or Dell'Orto 8mm
Power	1.25 horsepower
Weight (engine only) kilograms	7.8
Speed (1st gear) km/h	4–20
Speed (2nd gear) km/h	7–40

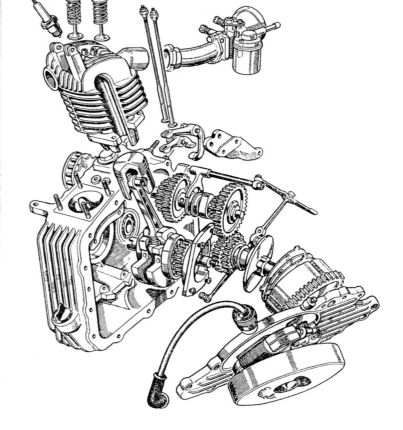

Exploded view of the T2. Pull rods still operated the two valves and an oil filler was positioned on the front of the crankcase.

CUCCIOLO T2, T50 1948–1952
Differing from the T1

Compression ratio	5.5:1 (6.5:1 Sport and T50)
Carburetion	Weber-Cucciolo 14/8
Power	0.8 (1.25 Sport) (1.5 T50) horsepower
Weight (engine only, kg)	8
Top speed (km/h)	35 (Turismo)

In 1950, the T2 was updated to the T50, with the oil filler at the rear of the crankcase.

CUCCIOLO T2

For 1948, recently appointed Chief Engineer Giovanni Fiorio redesigned the Cucciolo engine, creating the T2. Most of the developments were aimed at improving the ease of manufacture, noticeably the inclusion of a removable cylinder. The cylinder finning was horizontal, and the cylinder bolted via a four-stud flange to a new one-piece alloy crankcase. The revised cylinder head included a forward-facing exhaust, and the crankcase now included an oil filler at the rear of the engine.

Produced in two versions, the Turismo and Sport, the T2 was also available to special order as a higher performance T2 Sport that included a slightly larger piston (creating a full 50cc) and an 8.9:1 compression ratio to produce 2 horsepower at 5,700 rpm. When installed in a racing bicycle frame, it was capable of nearly 40 miles per hour. As early as February 1947, the Cucciolo was winning races in the *Micromotore* class, and the T2 Sport was the first of many special factory machines available for the privateer racer.

With sales booming, a more basic T0 was offered for 1949. With fixed speed and limited power, it proved unpopular and lasted only one year. An updated version of the T2 appeared in 1950, known as the T50. Apart from the crankcase oil filler moved from the front to the rear, there were updates to the clutch, primary drive, and carburetor.

CUCCIOLO T3

Another development of the Cucciolo T1 appeared during 1948, the T3. Fiorio bored the existing Cucciolo to 60cc, enclosed the exposed valves with grease lubrication, and installed a three-speed transmission. A foot gearshift was separate from the bicycle pedal cranks that were still employed for starting. This engine was to form the basis of the first complete Ducati motorcycle, but was initially displayed in an unusual three-wheeled machine, the Girino. This didn't make it into production, but in the meantime the T3 was prepared for racing by a number of private entrants. However, it was never as successful as the 50cc version because the 65cc Moto Guzzi Guzzino outclassed it.

CUCCIOLO T3 1948–1949
Differing from the T2

Bore (mm)	43.8
Displacement (cc)	60
Gears	Three-speed

60 1949–1950
Differing from the Cucciolo T3

Bore (mm)	42
Stroke (mm)	43
Displacement (cc)	59.57
Compression ratio	8:1
Carburetion	Weber Type 15 MFC
Power	2.25 horsepower at 5,000 rpm
Front suspension	Telescopic fork
Rear suspension	Cantilever
Weight (kg)	44.5
Colors	Dark red with gold pinstriping
Engine and frame numbers	From DM 60001

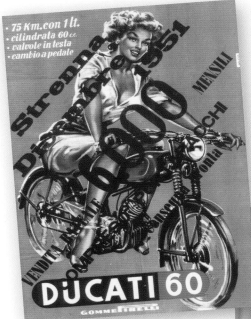

Although the first series 60 finished during 1950, it remained available through 1951 as this advertisement proclaims.

The first Ducati motorcycle was this 60 of 1949. The engine was a T3 with enclosed valves while the frame was by Caproni.

60

Faced with a proliferation of competition in the clip-on motor market, Giovanni Fiorio's 60cc T3 of 1948 led Ducati to consider the production of a complete motorcycle. Ducati had no experience in chassis manufacture, but many other manufacturers were developing special frames for the Cucciolo. One was a tubular steel frame with rear suspension by Caproni of Rovereto, a company primarily known for prewar airplane manufacture. Ducati subsequently asked Caproni to supply frames for the first Ducati, the 60. This first appeared in 1949, but by May 1950 Caproni decided to produce its own motorcycle and the relationship ended.

The 1949 60 was very advanced for its day, and markedly different to the second type, introduced in March 1950. The first engine was essentially a Cucciolo T3, with pull rods, a revised bore and stroke, a higher compression ratio, and a slightly larger Weber carburetor. The Caproni pressed-steel frame included a telescopic front fork with coil springing and a cantilever rear suspension with a spring under the seat and two friction dampers. The 60 was the first model to bear the engine and frame designation DM.

RACING CUCCIOLOS

Cucciolos were extremely successful in Italian street racing in the late 1940s. These are T2s racing during 1948. *Ducati*

As racing was endemic in Italy, it was inevitable that the Cucciolo would be adapted for competition in the *Micromotore* class for machines up to 50cc. In February 1947, Mario Recchia gave the Cucciolo its first victory, and for 1948 Ducati offered its first racing engine, a T2 Sport to special order. The T2 Sport was considerably successful during 1948, particularly in hillclimbs where its four-stroke torque provided an advantage over the two-stroke opposition, but also on street circuits. It wasn't until 1950 that Ducati produced its first complete motorcycle; all Cucciolo racers at this stage featured proprietary frames.

Ducati officially entered competition for the first time by assisting Ugo Tamarozzi in his attempt on a series of world records at Monza on March 5, 1950. The fifty-six-year-old Tamarozzi prepared the machine in his cellar in Milan, and with a 9.5:1 compression ratio and a 12mm Dell'Orto carburetor, and a 50/50 petrol/benzol fuel mixture, he scrupulously observed a 3,500-rpm limit to set six distance and six speed records. His fastest record was the 100 miles at 66.092 kilometers per hour. Not content with this, in May Tamarozzi returned with a co-rider Glauco Zitelli and a full factory machine prepared by chief engineer Giovanni Fiorio, setting twenty more records. The Cucciolo ran for twelve hours at 67.156 kilometers per hour. In November 1950 Tamarozzi and a team of riders set another twenty-seven long-distance world records, including the 3,000-kilometer at 66.320 kilometers per hour and the forty-eight-hour at 63.200 kilometers per hour. These records stood for nearly five years.

During 1951 the factory persevered with the Cucciolo and 65 Sport in competition and was rewarded with Franco Petrucci's third in the under-75cc class in the Milano-Taranto road race. The little engine was also suited to off-road competition. A factory team was entered in the ISDT held at the Valli Bergamasche in Italy that year, and Tamarozzi earned a silver medal.

The Ducati team on 65TSs in 1951. They were moderately effective in the under 75cc category in Italian road racing. *Ducati*

New for 1951 was the 60 Sport, the second type here with the new frame that included rear suspension. Now with an overhead-valve pushrod engine, this was Ducati's first sportbike. *Ducati*

60 SPORT, 65 SPORT

As soon as the 60 entered production, Fiorio had another engine on the drawing board, a 65cc engine with pushrod-operated overhead valves. Still ostensibly based on the Cucciolo, this engine design would last for nearly two decades. The extra displacement came from a larger bore, and the power increase through an improved combustion chamber and a larger carburetor. The cylinder head layout was unusual in that the intake was on the left and exhaust on the right. Starting was by a kick-start lever on the right.

The 60 Sport became the 65 Sport during 1952, but apart from the tank decals and color, it was unchanged. *Ducati*

60 SPORT 1950–1952, 65 SPORT 1953
Differing from the 60

Bore (mm)	44
Displacement (cc)	65.38
Compression ratio	8:1
Valve type	Overhead valve, pushrod
Carburetion	Weber Type 16 MFC or Dell'Orto MA16B
Power	2.5 horsepower at 5,500 rpm (5,600, 65 Sport)
Tires	1.75x22-inch
Rear suspension	Swingarm
Weight (kg)	48 (54, 65 Sport)
Top speed (km/h)	70
Colors	60S: Black frame, chrome tank with red panels
	65S: Dark red or green tank with chrome
Engine numbers	From 65T 06001

Initially the 65cc overhead valve engine was installed in the Caproni chassis of the 60, and called the 60 Sport. Soon after the demise of the Caproni connection, a new 60 Sport appeared, with the same engine in a new chassis. While the open pressed-steel frame and telescopic front fork was similar, the rear end included a swingarm with twin shock absorbers, allowing a rack to be installed behind the sprung solo saddle. The 60 Sport was renamed the 65 Sport during 1952, but was identical except for new colors.

In October 1951 Ducati Meccanica gained a new director, Dr. Giuseppe Montano. As a motorcycle enthusiast, Montano would have a profound effect on the direction of the company, immediately authorizing updated designs and approving an expanded racing program.

65T, 65TL

With the release of the 98 in 1952, the 65 was initially relegated to the role of mundane workhorse, initially as the touring 65T and higher specification 65TL (Turismo Lusso). These included a new pressed-steel open frame, similar to that of the 98, with a curved tubular section rear subframe. The rear rack was simpler, fitting over the rear fender. The 65 Sport was discontinued for 1954, but the 65T and 65TL continued until 1958.

"Basic transport for all"—that was how Ducati marketed the 65T in 1952. The Cucciolo engine origins are clearly evident. The frame was pressed steel and the suspension rudimentary. *Ducati*

65T, 65TL 1952–1954
Differing from the 65 Sport

Tires	2.00x22-inch Pirelli
Weight (kg)	54
Top speed (km/h)	70
Colors	**Mostly dark red**

BELOW: Mario Recchia rode this 1952 65T (engine number 06754) to work in Bologna during the week and raced it on Sundays. The 65T broke down in Sienna during the Milano-Taranto, Recchia dismantling and lubricating the bearings with a banana as he lacked oil. He finished the event.

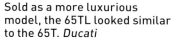

Sold as a more luxurious model, the 65TL looked similar to the 65T. *Ducati*

48 1952–1954
Differing from the Cucciolo T2

Carburetion	Weber 14mm
Power	1.5 horsepower at 5,500 rpm
Gears	Three-speed
Tires	1.75x24-inch Pirelli
Brakes	Double drum
Weight (kg)	41
Top speed (km/h)	50
Colors	Light gray, black highlighting

The 48 of 1952–1954 was more of a motorcycle than earlier Cucciolos and included front and rear suspension.

48

After Tamarozzi and Zitelli's successful speed records, the Cucciolo was given a new lease of life as the 48 in 1952. Sold as a complete motorcycle, not just a clip-on engine, the 48 combined features of the T2 and T3 in a pressed-steel rigid frame. The 48cc engine included the T3's enclosed valves with grease lubrication and three-speed transmission. The pressed-steel frame featured a sprung parallelogram front fork and incorporated a 4-liter gas tank, a tire pump sitting underneath. An improvement over the bicycle-type Cucciolos was the 48's double drum brakes. And although an unremarkable machine, the 48's primary appeal was its exceptional fuel economy, a claimed 90 kilometers per liter. Other variants of the 48 were based on a bicycle-style tubular frame, the Type G and Type DG, while another version was the unpopular and short-lived Ciclo scooter, with valanced fenders and engine panels.

CRUISER

By 1950, with Cucciolo sales declining and Italy under the spell of the scooter, Ducati decided it should enter the expanding scooter market. No expense was spared in the Cruiser's development, and not content to build a Vespa or Lambretta two-stroke clone, Ducati had Giovanni Fiorio design a new four-stroke engine. Ducati also commissioned an outside design company to style the bodywork. Amazingly ambitious, it was the first four-stroke scooter and the first motorcycle with an automatic transmission. Envisaged as a luxury model appealing to a new, and more prosperous, clientele, the Cruiser was decades ahead of its time.

Some of the Cruiser's engine design features came from the overhead-valve 65 and a single Dell'Orto carburetor mounted directly on the cylinder head, feeding downdraft into the cylinder. Initially the engine produced 12 horsepower, but was later detuned to 7.5 horsepower due to a government-imposed 50-mile-per-hour speed limit for scooters. A first for a scooter was the standard electric starter, and at a time when just about all cars and motorcycles used a weak 6-volt electrical system, the Cruiser had 12-volt electrics. The 45-watt dynamo powered a huge 32-amp hour battery, providing exceptional lighting.

ABOVE LEFT: Although advanced for the day, the Cruiser was heavy and complicated, and it was an unsuccessful attempt by Ducati to break into the scooter market.

ABOVE: Until mid-1956, this SSR logo was still incorporated on brochures and on some motorcycles.

Even more remarkable than the engine and electrical system was the gearbox. Running longitudinally under the seat, this was automatic, with a hydraulic torque converter housed in an alloy casting that incorporated the swingarm pivot. Unfortunately, while commendable in endeavoring to make the Cruiser a user-friendly machine, the automatic gearbox was extremely complicated and problems resulted in numerous warranty claims.

The rear suspension was unusual in that it comprised a long arm connected to a cylinder with rubber inserts for damping, the unit positioned horizontally underneath the engine. The heavy swingarm/transmission casting worked like a fluctuating arm. The front suspension was a patented new type, similar to the swinging shackle scooter type, but with a single hydraulic shock absorber.

Although Ducati initially only admitted that the bodywork was designed "by a well-known car design company," it eventually transpired that it was completed by Ghia, a company better known for designing luxury cars. The Cruiser's styling was unremarkable, but when released

CRUISER 1952–1954

Engine	Single-cylinder four-stroke, air-cooled
Bore (mm)	62
Stroke (mm)	58
Displacement (cc)	175
Compression ratio	7.5:1
Valve type	Overhead-valve, pushrod
Carburetion	Dell'Orto
Power	7.5 horsepower at 5,600 rpm
Gears	Automatic
Tires	2.45x10-inch Pirelli
Weight (kg)	154
Top speed (km/h)	80
Colors	Blue and gray, red and gray
Number produced	2,000

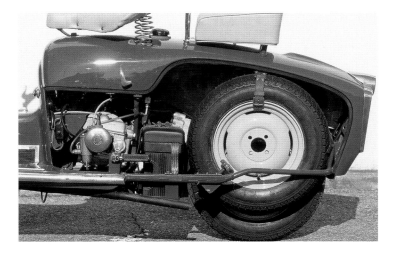

A spare tire was fitted behind the engine, along with a large battery.

at the Milan Show in January 1952, the Cruiser was initially hailed as a highlight because of its technical innovation. However, with no reputation for manufacturing scooters, against the established and successful Vespa and Lambretta, the Cruiser was seen as heavy and complicated. Soon earning a reputation for unreliability, the Cruiser was doomed. Although the model lasted until 1954, Ducati still didn't learn from this experience, repeating the scooter debacle a decade later.

The first 98 was rather dull in its gray livery, but finally broke away from the Cucciolo origins. This is a 1952 example, number 20402, but with a double-bolt seat mount.

98

Alongside the Cruiser at the 1952 Milan Show was a new motorcycle, the 98. Also designed by Fiorio, this was a development of the 65, but it was more conventional in its execution. The overhead valve four-stroke engine had a forward-facing exhaust and rear carburetor, with an external oil pipe on the right to feed the valve gear. The cylinder was inclined 25 degrees, and the two valves were set at an included angle of 100 degrees. The engine cases were restyled to smoothly incorporate the gearbox, which was initially still three-speed. The dynamo was also enclosed, in a vented cover, and the kick-start was on the left.

Also new was the open spine-type pressed-steel frame, and with the heavy valanced fenders and substantial, but undamped, telescopic fork, the 98 looked more like a proper motorcycle than the 65. A small (10-liter) gas tank sat atop the frame spine, and on the early versions a single fastener supported the rear fender and rack. This soon changed, with a longer rear subframe incorporating an additional mount. Apart from a different muffler and colors, the 98 continued virtually unchanged into 1954.

98 1952–1954
Differing from the 65

Bore (mm)	49
Stroke (mm)	52
Displacement (cc)	98.058
Compression ratio	8:1
Carburetion	Dell'Orto MA16B
Power	5.5 horsepower at 6,800 rpm
Tires	2.75x17-inch Pirelli or CEAT
Front suspension	Telescopic fork
Rear suspension	Swingarm
Wheelbase (mm)	1,240
Weight (kg)	72
Top speed (km/h)	75
Color	Gray
Engine and frame numbers	From DM 10001

New for 1953 was the second version of the 98, the 98T. This had a dual seat, larger fuel tank, and new colors. *Ducati*

Also new for 1953 was the 98TL with large engine protection bar. *Ducati*

98T, 98TL 1953–1954
Differing from the 98

Carburetion	Dell'Orto MA18B (98TL)
Power	5.8 horsepower at 7,500 rpm
Tires	2.50x17-inch, 2.75x17-inch
Wheelbase (mm)	1,245
Weight (kg)	80
Top speed (km/h)	75
Engine and frame numbers	From DM 1001

98T, 98TL

During 1953 Ducati was split into two divisions: Ducati Meccanica, responsible for motorcycle production, and Ducati Elettrotecnica, the electrical division. Although operating independently, both were under the parent SSR company until 1956, and in the meantime, in order to increase production, Montano instigated modernization of the Borgo Panigale factory. As a result, more variations on the 98 appeared for 1953.

Joining the 98 was the 98T (Turismo) and 98TL (Turismo Lusso). The 98T received a restyled larger gas tank (now 14 liters), new handlebar, dual seat, and new colors, but the engine and running gear was identical to the earlier 98. The 98TL was based on the 98T, but with light alloy wheel rims, a higher braced handlebar, a different tank, and a two-piece saddle. The 98TL also had a large protective crash bar and a larger carburetor.

98 SPORT, 98 SUPER SPORT

The first sporting Ducati also appeared at the Milan Show in early 1953, the 98 Sport. Along with a four-speed transmission, the engine was uprated, the larger carburetor mounted on a longer intake, and attached to the front of the sump was a finned oil cooler. The gas tank

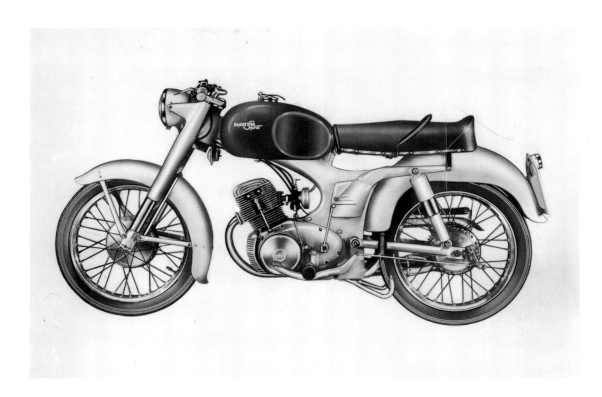

The 98 Sport was Ducati's first sporting model and now included a four-speed transmission. The first Sport rear subframe included only one seat-bolt mount. *Ducati*

was shared with the 98TL, but the handlebars were lower, and the 17-inch wheels with light alloy rims included a full-width drum front brake. The Sport was an interesting variation on what had been a range of mundane motorcycles up until that date, but the 98 Super Sport that appeared in 1954 was even more so. A small handlebar-mounted fairing and screen augmented the sculptured 14-liter gas tank. The engine was further uprated, and the suspension new. Because of the success of the overhead camshaft Marianna, the 98 Super Sport only lasted until 1955, but the 98 Super Sport chassis and styling was inherited by the 98 Sport from 1956. The 98 Sport and 98 Super Sport were certainly the most attractive offerings of all the rather dull overhead valve models.

The 1954 98SS included a new front fork and 20mm carburetor. *Ducati*

98 SPORT 1953–1954, 98 SS 1954–1955
Differing from the 98 and 98N

Compression ratio	9:1 (10:1)
Carburetion	Dell'Orto MA18B (MB20B)
Power	6.8 (7) horsepower at 7,300 rpm
Gears	Four-speed
Tires	2.50x17-inch, 2.75x17-inch CEAT
Brakes (mm)	158 front, 135 rear
Weight (kg)	75
Top speed (km/h)	90 (95; 98 SS)
Color	Black/silver
Engine and frame numbers	Until DM 06000

With the release of the overhead valve 98 in 1952, Ducati decided to concentrate on the 100cc racing category but had little success until Taglioni supervised their preparation for the 1954 ISDT held in Wales. Although the smallest machines entered in the event, Alberto Farnè and Giovanni Malaguti both won silver medals. In the 1954 Motogiro d'Italia, Ducati entered five 98s—ridden by Alberto Gandossi, Silvio Landi, Sergio Saccomandi, Mario Recchia, and the young Franco Farnè. But despite Gandossi winning two sections and finishing third overall, the Laverdas humbled the Ducati team. Montano was displeased and immediately enticed Fabio Taglioni from Mondial with a one-year contract to turn Ducati's racing fortune. This would be the turning point for the company.

The Ducati stand at the 1954 Milan Show. In the foreground is the 98 Sport.

CHAPTER 2
TAGLIONI'S ARRIVAL
THE GRAN SPORT AND DESMODROMICS OVERHEAD CAMSHAFT AND OVERHEAD-VALVE SINGLES, 1955–1960

Fabio Taglioni was Ducati's chief engineer for thirty-five years and the father of desmodromic valve operation. *Ducati*

During the 1950s most Italian smaller displacement motorcycle manufacturers strived for good results in the Motogiro and Milano-Taranto Gran Fondo road races. Of particular importance to Ducati was the Motogiro d'Italia, for motorcycles less than 175cc and held over nine days in stages held on normal Italian roads. Success in the Gran Fondo events was considered pivotal for sales, and only one month after the 1954 Motogiro humiliation, Ducati's managing director Giuseppe Montano lured Fabio Taglioni from Mondial, commissioning him to design a new motorcycle capable of winning the 1955 Motogiro.

The result was the magnificent Gran Sport, later nicknamed the Marianna. This advanced design formed the basis of the all the overhead camshaft singles through until 1975, and many of its design characteristics still are featured on some current engines. After the humiliation of 1954, no one could have predicted the success of the Marianna in the 1955 Motogiro d'Italia. The Mariannas were unbeatable, initiating a racing record that continues today. Taglioni's one-year contract was extended, and his association with Ducati would last four decades.

Fabio Taglioni with Battilani, Vighi, Villa, and Lelli at the release of the Gran Sport at Imola, prior to the 1955 Giro d'Italia. *Ducati*

FABIO TAGLIONI

When he came to Ducati on May 1, 1954, the thirty-three-year-old Fabio Taglioni already had a considerable reputation as a motorcycle engineer. Born on September 10, 1920, Taglioni hailed from Lugo, in the center of the Emilia Romagna region, an area known for its rich automotive engineering tradition. Taglioni's interest in motorcycles began during the 1930s, and after sustaining a leg wound during World War II, he completed his engineering studies at Bologna University in three years instead of the usual five by teaching himself. Graduating in 1948, he taught at a technical college in Imola, and with the help of his pupils he designed a racing 75, selling this to the small Bolognese motorcycle company Ceccato. This eventually led to his appointment as an assistant to Alfonso Drusiani at FB Mondial in 1952, an experience that would shape Taglioni's future. Mondial had just won three consecutive 125cc World Championships and Drusiani's double overhead camshaft racing 125 served as an inspiration for Taglioni when he came to design the Gran Sport.

Taglioni's success with the desmodromic Grand Prix racers provided him with an international reputation, but he always remained loyal to Ducati, resisting the temptation of lucrative offers to work in the automobile industry. Although a prolific designer, with more than 1,000 drawings to his credit, Taglioni was primarily interested in racing, often to the detriment of the production models. His formula for a racing machine called for, "light weight, simplicity, a narrow engine, and a wide power band."

What was most impressive about Fabio Taglioni was his humility. He was never arrogant or dogmatic, and always acknowledged the support of others. Although sometimes criticized for his conservatism and reluctance to adopt new technology, Taglioni remains the premier figure in the history of Ducati racing. He was the father of desmodromic valve operation, and the only one to make it work successfully on a motorcycle. Even after his official retirement in 1989, Fabio Taglioni was involved in engine development, later in life suffering emphysema, the result of a lifetime of smoking. Until his death on July 19, 2001, he remained a revered figure at Borgo Panigale, and every time a desmodromic Ducati wins a race his legacy continues.

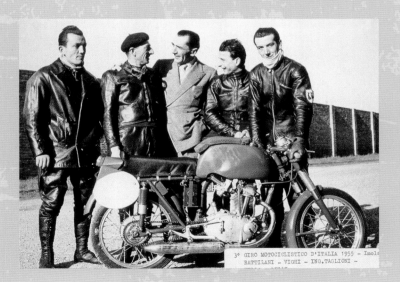

65TS 1955–1958	Differing from the 65 Sport
Compression ratio	8:1
Tires	2.00x18-inch CEAT
Colors	Mostly red/black
Frame numbers	From DM 50001

New for 1955 was the 65TS, and in the wake of Ducati's Motogiro success, this was very popular.

1955

While Taglioni was working on the Gran Sport, the existing range of Cucciolos and overhead valve singles continued, now including the 65TS, an updated 98 Sport, and the final Cucciolo, the 55.

65TS, 65T, 65TL

While the 65T and 65TL continued unchanged, new this year was the 65TS (Turismo Sport), similar to the earlier 65 Sport, but with the T and TL chassis. Along with a dual seat, the gas tank was reshaped, and a small screen complemented the low handlebars. This was an extremely successful model, coinciding with Ducati's success in the Motogiro d'Italia that year, and it was the best-selling motorcycle in Italy during 1955. It lasted until 1958.

Another updated model was the 98 Sport. All 98 Sports featured the front-mounted oil cooler.

98 Sport, 98T, 98TL, 98

A new 98 Sport was also available for 1955, the pressed-steel chassis and styling inherited from the earlier 98 Super Sport. The earlier 98, 98T, and 98TL continued unchanged this year.

55E, 55R

Replacing the 48 for 1955 was the final development of the Cucciolo, the M55. Still displacing 48cc, and remarkably similar to Farinelli's near decade-old design, the M55 was installed in a new pressed-steel frame with moped-style leading link forks. The 55E (*Elastico*) had rear suspension, while the 55R (*Rigido*) had a rigid rear end. In some respects the M55 was lower specification than the 48, and although it featured a more efficient silencer instead of an expansion box, the engine made less power, and the transmission was only two-speed. As with the earlier Cucciolo, starting was by bicycle pedals.

By the time the M55 was released, the motorcycle world was changing. As a more affluent clientele moved toward affordable automobiles, motorcycle sales began to decline in the mid-1950s. Cheap, basic motorcycles for transportation, such as the M55, suffered most, and sales never met expectations. By 1957, the Cucciolo had had its day, but with more than 400,000 produced, it put Ducati on the map as a motorcycle manufacturer.

98 SPORT 1955–1958
Differing from 1954

Compression ratio	10:1
Carburetion	Dell'Orto MB20B
Power	6.5 horsepower at 7,000 rpm
Color	Gray, gray/black
Engine and frame numbers	From DM 06001

Now with front and optional rear suspension, the M55 was the final version of the Cucciolo.

55E, 55R 1955–1957
Differing from the 48

Compression ratio	6.7:1
Carburetion	Weber Type 14 MFC
Power	1.35 horsepower at 5,500 rpm
Gears	Two-speed
Tires	2.00x18-inch CEAT
Weight (kg)	45
Top speed (km/h)	48
Colors	Light gray or red

With its external flywheel and narrow crankcase, the M55 engine retained strong links to the original Cucciolo.

TAGLIONI'S ARRIVAL

The 100cc Gran Sport established Ducati's racing tradition.

100 Gran Sport (Marianna)

As it was always envisaged for racing, Taglioni's Gran Sport represented a significant departure for Ducati from the existing overhead-valve singles and would also influence Ducati engine design for the next six decades. The basis was a vertically split aluminum unit construction sand-cast crankcase, the cylinder inclined forward 10 degrees with the single overhead camshaft driven by a set of straight-cut bevel gears. In an era where nearly all the competition featured pushrod valve operation, this immediately gave the Gran Sport an advantage and was a reflection on the quality of machinery just installed at the Ducati factory. All the bearings were ball or roller, a feature that would carry through on all racing Ducati engines until the 1979 Pantah.

A four-speed gearbox was incorporated in the crankcases, driven by straight-cut primary gears with a wet multiplate clutch. Ignition was by battery and coil, and the cylinder head was in two pieces. The valves were opposed at 80 degrees and closed by exposed hairpin valve springs. This engine was placed in a single downtube tubular steel frame that utilized it as a stressed member, another feature that would continue for decades on many Ducatis. Suspension was by a telescopic fork and twin shock absorbers, and the brakes were magnesium from Amadori. A generator, horn, and headlamp were fitted to comply with Italian FMI regulations, and instead of an air pump, a compressed air canister sat on the left rear frame tube.

With the model officially unveiled on March 5, 1955, thirty-seven Gran Sports were entered in the 3,400-kilometer Motogiro d'Italia. Riders included Gianni Degli Antoni, Leopoldo Tartarini, Francesco Villa, Antonio Graziano, Ettore Scamandri, Franco Farnè, and Giuliano Maoggi. After the disappointment of the previous year, no one could have predicted their success, Mariannas winning every inter-city stage. Degli Antoni easily won the 100cc class at an average speed of 98.90 kilometers an hour, good enough for fifth overall. Not only was Taglioni's future assured, Ducati was able to advertise itself as "the miracle of the Motogiro." The Marianna became available as a cataloged model but was always an expensive limited-edition production racer.

LEFT: Degli Antoni was Ducati's most successful rider in the 1955 Motogiro. *Ducati*

ABOVE: Aldo Forcinelli was one of thirty-seven riders entered on a Ducati in the 1955 Motogiro. *Ducati*

BELOW: Leopoldo Tartarini with the Gran Sport in the 1955 Motogiro d'Italia. *Ducati*

100 GRAN SPORT 1955–1957

Engine	Single-cylinder four-stroke, air-cooled
Bore (mm)	49.4
Stroke (mm)	52
Displacement (cc)	99.66
Compression ratio	8.5:1
Valve actuation	Bevel gear–driven single overhead camshaft
Carburetion	Dell'Orto SSI 20C
Power	9 at 9,000 rpm (9.7 at 9,800 rpm to 12 at 10,500 rpm during 1956)
Gears	Four-speed
Front suspension	Telescopic fork
Rear suspension	Twin shock absorber swingarm
Tires	2.75x17 inch
Brakes	160mm front and rear
Wheelbase	1,280mm
Dry weight	80 kg
Top speed (km/h)	130
Engine numbers	From 001

TAGLIONI'S ARRIVAL | 27

The 125 Gran Sport was introduced shortly after the 100. While it was very similar, the wheels were larger, 18-inch instead of 17-inch.

125 Gran Sport

With the Milano-Taranto race looming in June, Taglioni prepared a 125cc Marianna for Maoggi. The engine was bored to slightly, the valves enlarged, and a larger Dell'Orto carburetor fitted. Despite the improved power and larger diameter wheels, it wasn't enough for Maoggi to beat his teammate Degli Antoni on the 100 Marianna in the Milano-Taranto this year.

125 GRAN SPORT 1955–1957
Differing from the 100 Gran Sport

Bore	55.25mm
Capacity	124.8cc
Carburetor	Dell'Orto SSI 20C or 22C
Horsepower	12 horsepower at 9,800 rpm (14 at 10,000 rpm from 1956)
Wheels	18-inch front and rear
Dry weight	85 kg
Engine numbers	From 501

The single overhead camshaft engine was also similar, but carried specific "125" stamps.

1956

As the debut racing season proved even more successful than anticipated, Taglioni was allowed to develop the Gran Sport with a double overhead camshaft cylinder head for 1956, the Bialbero. But as success was still vital in the Motogiro and Milano-Taranto, the single overhead camshaft Marianna was refined and the team managed to better its 1955 results. Maoggi went on to win the event outright on his 125, again the 100cc Marianna swamping the smaller category, Gandossi leading home Villa, Bruno Spaggiari, Scamandri, and Farnè. Mariannas also took out every stage of the race.

Results were similar in the final Milano-Taranto race, held on June 10. Here Degli Antoni astounded everyone by finishing sixth overall on a 125 Marianna at 103.176 kilometers an hour, ahead of most of the 500s and 250s. The dominance of the Marianna continued in domestic events during 1956, Franco Farnè winning the 100cc Junior Italian Championship. Later in the year the Gran Sport engine grew to 175cc, but this wasn't very successful, the 175 F3 eventually replacing it.

125 Grand Prix (Bialbero)

While Ducati's victories in the MSDS (*macchine sport derivate dalla serie*) category for production sports machines were impressive, Montano and Taglioni wanted to win the modified sports machine class. For this, Taglioni created the Bialbero, or double camshaft 125 Grand Prix, this officially unveiled on February 25, 1956.

Essentially a Gran Sport apart from the cylinder head, the twin overhead camshaft layout immediately provided 15.5 horsepower at 10,500 rpm. Taglioni soon managed to raise the power and maximum revs slightly. Early examples featured a Marianna-derived two-piece cylinder head, but this was problematic with the relatively high compression ratio and high lift camshafts. Even with the head and cam box cast together, the Bialbero was still considerably down on power compared to the works MVs. With a handlebar fairing and streamlined seat, the top speed was around 170 kilometers an hour. Taglioni also included a fifth gear outside the crankcase wall, behind the clutch inside the primary drive case. The basic chassis was that of a Marianna and included a dustbin fairing for 1956, while the magnesium Amadori brakes featured large forward scoops. Even in the hands of Degli Antoni and veteran Alano Montanari, the early Bialbero was outclassed in 1956 Italian Championship events. Although Taglioni turned to Desmodromics on the factory machines, the Bialbero continued to be developed alongside the Desmo and was made available to privateers in limited quantities.

Ducati's success in the Motogiro continued for 1956. This is Antonio Graziano in the snow on the Pian delle Fugazze. *Ducati*

The first Bialbero was very similar to the Gran Sport and retained a two-piece cylinder head.

125 GRAND PRIX BIALBERO 1956–1959

Bore (mm)	55.25
Stroke (mm)	52
Compression Ratio	9.8:1 (1956), 11:1 (from 1958)
Capacity (cc)	124.6
Valve actuation	Bevel gear–driven double overhead camshaft
Carburetor	Dell'Orto SSI 26 or 27mm (22mm for F2)
Horsepower	16 at 11,500 rpm (17 at 10,500 rpm during 1958)
Gearbox	Five-speed
Wheels	18-inch front and rear (17-inch from 1958)
Tires	2.50x18 and 2.75x18 (2.25x17 from 1958)
Brakes	180x40 and 160x40 (180x25 front and rear from 1958)
Wheelbase (mm)	1,219
Dry weight (kg)	90

Desmodromic 125

Although the 125 Bialbero offered improved performance over the Marianna, its reliability was questionable at high rpm. Valve-to-piston clearance was critical with the higher compression ratios now required, compounded by the problem of valve float at higher rpm. During his days at university, Taglioni had shown interest in desmodromic, or positive valve actuation, and almost as soon as he joined the company, he began to work on a desmodromic cylinder head for the Gran Sport. During 1955, he produced the first desmodromic prototype, Taglioni's faith in the concept reassured by the success of the Desmodromic Mercedes W196 Grand Prix and 300 SLR sports cars during 1954 and 1955.

Taglioni's design included triple overhead camshafts (Trialbero), with two outside opening camshafts (like the Bialbero), and a central camshaft closing the valves through forked rockers. The cylinder head design was similar to the Bialbero, and with a 27mm Dell'Orto, the initial power output was only marginally increased over the Bialbero, but with a much higher safety margin. The little Desmo could be run to 14,000 rpm on the overrun. Below the cylinder head, the

ABOVE: The early factory Desmo racers carried the prancing horse symbol of Italian World War I fighter ace Baracca. *Ducati*

LEFT: The 1956 125 Desmo chassis was also based on the Marianna. *Ducati*

BELOW: Degli Antoni was the victor at Hedemora in Sweden in 1956. *Ducati*

early desmodromic engine was virtually identical to the Bialbero, with a five-speed gearbox and installed in a similar chassis.

Ducati's goal at the beginning of 1956 was to acquit itself at the Nations Grand Prix at Monza in September. So in preparation for this important event, the Desmo's debut took place unannounced at the nonchampionship Swedish Grand Prix at Hedemora on July 15, 1956. In the hands of Taglioni's star rider Degli Antoni, the 125 Desmo won the 125cc race at an average speed of 84.45 miles per hour, soundly beating a host of private MVs and Mondials.

Unfortunately, disaster struck shortly afterward. During testing at Monza prior to the Nations Grand Prix on August 7, 1956, Degli Antoni lost control rounding the Lesmo curve and died. Although Alberto Gandossi was drafted in as a replacement, the pall cast by Degli Antoni's death proved too much, and the three Desmos were swamped by Ubbiali's MV in the race. Although the Desmo would be raced in Italian events during 1957, Ducati decided to wait until 1958 before seriously attempting Grands Prix.

RIGHT: The first desmodromic 125 engine was still based on the Marianna. *Ducati*

125 DESMO 1956–1960
Differing from the 125 Grand Prix

Compression Ratio	10:1 (1956), 10.2:1 (from 1958)
Valve actuation	Bevel gear–driven desmodromic triple overhead camshaft
Carburetor	Dell'Orto SSI 27 or 29mm
Horsepower	17 at 12,500 rpm (1956) to 21.8 at 11,800 rpm (1960)
Gears	Five- or six-speed
Wheels	18-inch front and rear
Tires	2.50x18 and 2.75x18
Wheelbase (mm)	1,270
Dry weight (kg)	80

SILURO (TORPEDO)

As setting speed and distance records were extremely popular methods for publicity in the 1950s, two Milanese riders, Mario Carini and Santo Ciceri, encouraged Ducati to support them in a record attempt at Monza on November 30, 1956. Ducati agreed to prepare the engine and commission aluminum streamlining from the Milanese company Nardi and Danese. With the aid of a wind tunnel, the design featured an open cockpit and a small Plexiglas screen, and it was extremely effective. Supported by additional tubing, underneath the aluminum bodywork was virtually a standard 100cc Marianna. Despite the threat of rain, Carini and Ciceri set a total of forty-four new records, with the 100cc Marianna timed at 171.910 kilometers per hour.

Carini and Ciceri's streamlined 100cc Siluro at Monza in Novemeber 1956. They set a total of forty-four new speed and distance records. *Ducati*

PRODUCTION MODELS 1956

Although the racing overhead camshaft singles were stealing the limelight, production continued to expand at Borgo Panigale, the Marianna's success seeing Ducati growing to become one of the largest motorcycle manufacturers in Italy. During 1956, Ducati sold 10,767 motorcycles (3.5 percent of the market), and employees numbered more than 700. These were halcyon days, and it wouldn't be until 1992 that these sales would be matched, but the range was still confined to the less exciting overhead camshaft singles.

This year the overhead-valve engine grew to 125cc, in the 125TV and 125T, and alongside the two 125s were four 98s, three 65s, and two 55s.

125TV, 125S, 125T

Little externally distinguished the early 125 engine from the 98 of the same era, and it still incorporated the external oil line on the right and finned cylinder head without a separate rocker cover. The 125TV and 125S were the first models to feature the tubular-steel double downtube frame, and along with upgraded suspension and a full-width front brake, they promised improved road behavior. A period styling touch on the TV was the nacelle over the front fork incorporating the headlight. The 125S had clip-on handlebars while the 125T was an economy model, the tubular-steel spine frame without downtubes and a less sophisticated suspension and smaller brakes. Two types were available, one with a solo seat and rack, and the other with a dual seat (*sella lunga*).

The 125TV was the production range leader for 1956.

125TV, 125S, 125T 1956–1960
Differing from the 98

Bore (mm)	55.2
Displacement (cc)	124.443
Compression ratio	7:1
Carburetion	Dell'Orto ME20B (MA18B, 125T)
Power	6.5 horsepower at 6,500 rpm
Gears	Four-speed
Tires	2.50x17-inch, 2.75x17-inch
Weight (kg)	95
Top speed (km/h)	86
Colors	Red
Engine and frame numbers	From DM 1001 (125TV,T) From DM125S 4000 (125S)

ABOVE: With a frame without front downtubes and more basic suspension, the 125T was a budget model. *Ducati*

BELOW: The 98T received a new frame for 1956 but retained the three-speed transmission. *Ducati*

98TL, 98S, 98T, 98N, 65TS, 65T, 65TL, 55E, 55R

For 1956, the 98TL was restyled along the lines of the new 125, sharing the same tubular steel frame, but without the double front downtubes. The 98TL inherited the suspension and running gear of the earlier 98SS but had steel wheel rims. The engine specification was unchanged. Also sharing the new tubular steel frame was the 98T, now with a solo seat and rack. This was now a base model alongside the pressed-steel frame 98N, this year replacing the 98, also with a solo seat, and still with a three-speed gearbox. The 98 Sport continued largely unchanged, inheriting the earlier 98SS pressed-steel chassis and styling, while the three 65s and two 55s also continued as before.

TAGLIONI'S ARRIVAL

1957

At the end of 1956, Ducati released its first production overhead camshaft single, the 175T, followed soon afterward by the 175 Sport. The 175's release coincided with a boom in motorcycle sales, with more than three million motorcycles over 50cc sold in Italy during 1956. The immediate success of the 175 Sport and 175T (they accounted for 25 percent of 175cc sales in Italy during 1957) saw a proliferation of overhead camshaft models during the next few years. Another significant occurrence that would shape Ducati's future for the next decade and beyond was the appointment of Berliner as the US distributor, and Berliner's influence would ensure Ducati would survive the sales downturn of the late 1950s and early 1960s.

With the Milano-Taranto canceled following the tragedy during the Mille Miglia in May, the 1957 Motogiro d'Italia was the final Italian road race. Again the Mariannas dominated, winning every 100cc stage and eight 125 stages. Mandolini won the 100cc category and Graziano the 125. Even with the end of the Gran Fondo races, the Marianna was still raced with success, notably at the Montjuïc 24-hour race at Barcelona in July 1957. On a 125 intended for the Milano-Taranto, Spaggiari and Gandossi won the event outright at an average speed of 57.66 miles per hour.

Taglioni also designed a 175 twin, this displayed at the Milan Show at the end of 1956. Leopoldo Tartarini rode this in the 1957 Motogiro d'Italia, but retired with ignition and generator problems during the third stage. With its twin overhead camshafts driven by a train of spur gears from a jackshaft between the cylinders, the 175 formed the basis for all the later racing parallel twins. Complex and difficult to work on, the engines were beautifully constructed, with the flywheels and big-end assemblies machined from solid steel and all the gears drilled for lightness. The clutch was dry,

98N 1956–1957 Differing from the 98

Gears	Three-speed (four-speed from 1957)
Weight (kg)	81
Top speed (km/h)	80

ABOVE: The 98TL production line in 1956. This year a new tubular-steel frame replaced the earlier pressed-steel type. *Ducati*

Leopoldo Tartarini rode this 175 parallel twin in the 1957 Motogiro but retired. The twin was never as successful as the singles. *Ducati*

Marcello Sestini on a Gran Sport in the 1957 Motogiro d'Italia, the final Gran Fondo road race of the 1950s. *Ducati*

the hairpin valve springs exposed, and with an 11:1 compression ratio and 18mm Dell'Orto carburetors, the 49x46.6mm 175 produced 22 horsepower at 11,000 rpm. But at 112 kilograms, the 175 twin was too heavy and suffered in comparison to the single.

Grand Prix (Bialbero) and Desmodromic 125

For 1957, the Bialbero received a new cylinder head casting with a distinctive polished alloy gear cover and an additional oil supply direct to the camshafts. On some examples, the frame now incorporated developments from the Desmodromic 125, including a strengthening dual loop subframe under the engine.

During 1957, Taglioni managed to increase the power of the 125 Desmo to 19 horsepower at 13,000 rpm and also produced a 100cc version for Italian Formula 2 racing. In the hands of Franco Farnè, the 100 Desmo dominated the 1957 Italian Junior Championship. A new duplex cradle frame was also developed, and the Nations Grand Prix at Monza was the only Grand Prix outing for the Desmos this year. Again it was a disappointment as on the first lap Gandossi fell, bringing down a third of the field in the process.

LEFT: An improved Bialbero was available for 1957, the frame with additional strengthening underneath the engine and a new cylinder head.

RIGHT: The bevel-gear cover was now a beautiful one-piece polished aluminum casting.

The 175 Sport was the first sporting overhead camshaft single.

PRODUCTION MODELS 1957

175 Sport

The 175cc overhead camshaft engine established the blueprint for all Ducati production overhead camshaft singles until their demise in 1974. Although the aluminum overhead camshaft engine was based on the Gran Sport, the crankcases were now die-cast and the cylinder head was in one piece. To keep the engine oil tight, the hairpin valves springs were enclosed, and to reduce noise, the bevel gears and primary drive were helically cut.

Although the general engine layout followed the 100 and 125cc Gran Sport, the overall dimensions were increased slightly to accommodate the larger capacity and allow for a future increase to 200cc. The crankcase design was changed from the Gran Sport to incorporate a narrower front engine mount and wider rear mount. In the process of adapting the Gran Sport engine into production, a kick-start was added and this rudimentary addition would continue to be problematic on all narrow-case singles. Two Silentium exhaust systems were offered, with either a single or dual muffler.

The chassis was also based on that of the Gran Sport, with a single downtube frame that utilized the engine as a stressed member. As the 175 Sport was the first production overhead camshaft single, the 1957 version included a number of unique features, notably a bolted-on rear frame section and the front downtube consisting of two welded sections. The front engine mount only used two bolts. The first series (Type A) frames also included the swingarm bushes in the frame rather than the swingarm, resulting in a slightly narrower swingarm than the later Type B. The seven-fin Grimeca brakes would be featured on all 175 Sports, and later on some 200s and 250s.

175 SPORT 1957

Engine	Single-cylinder four-stroke, air-cooled
Bore (mm)	62
Stroke (mm)	57.8
Displacement (cc)	174.5
Compression ratio	8:1
Valve actuation	Bevel gear–driven single overhead camshaft
Carburetion	Dell'Orto MB22.5B
Power	14 horsepower at 8,000 rpm
Gears	Four-speed
Front suspension	Gualandi, Marzocchi, or Ducati fork
Rear suspension	Marzocchi twin shock swingarm
Brakes (mm)	Drum 180 front, 160 rear
Wheels	18x2.25-inch front and rear
Tires	2.50x18, 2.75x18 CEAT
Wheelbase (mm)	1,320
Dry weight (kg)	106
Top speed (km/h)	130
Colors	Copper and red
Engine and frame numbers	From DM175 75001

The 1957 version included a number of unique features, notably the frame with a bolt-on rear subframe and two-piece front downtube.

175T

The 175T's engine was ostensibly shared with the 175 Sport, including the crankshaft, con rod, and crankcases, but carburetion was by a slightly smaller Dell'Orto, and the four-ring Borgo piston included a lower compression ratio. The 175T frame was as on the 175 Sport, with the same bolted rear section and front engine mount. The 175T received a more traditionally styled fuel tank and deeper fenders, the very first with 17-inch wheels with steel rims, but these were soon 18-inch.

After an unsuccessful ride on the prototype 175 parallel twin in the 1957 Giro d'Italia, Leopoldo Tartarini decided to undertake a round-the-world trip on the new 175T. Teaming with a friend from school days, Giorgio Monetti, the pair left Bologna in September 1957, Ducati's boss Giuseppe Montano providing a champagne sendoff and technical support for the 60,000-kilometer journey. In what was a grueling endeavor at the time, the pair visited thirty-six countries in five continents, returning to a hero's welcome in Bologna a year after they left. Ultimately, the trip proved a hugely successful promotional exercise for Ducati, establishing new markets in many countries that had previously not been on Ducati's sales map.

175T 1957 Differing from the 175 Sport

Compression ratio	7:1
Carburetion	Dell'Orto MB22B
Power	12 horsepower at 6,000 rpm
Front suspension	Marzocchi 32mm fork
Brakes (mm)	Drum 158 front, 136 rear
Wheels	17-inch; 18x2-inch and 18x2.25-inch
Tires	3.00x17; 2.50x18 and 2.75x18 CEAT
Dry weight (kg)	104
Top speed (km/h)	110
Colors	Dark red with white or black highlighting
Engine and frame numbers	From DM175 00001

The 175T was the touring version of the overhead camshaft single. Early examples shared the same frame as the Sport.

ABOVE LEFT: Nineteen fifty-seven was the final year for the 125 frame without front downtubes. This is the Sella Lunga version. *Ducati*

ABOVE RIGHT: The 1957 65TS: still with the pressed-steel frame and rudimentary front suspension. *Ducati*

125TV, 125S, 125T, 98S, 98TL, 98T, 98N, 65T, 65TL, 65TS

With most resources directed to toward the introduction of the overhead camshaft 175 Sport and 175T, the existing range of overhead valve singles continued for 1957 with few changes. All 98s now shared the 98TL's tubular-steel frame and gained a four-speed transmission.

1958

This year saw Ducati mount a serious challenge for the 125cc World Championship, while the production range of both overhead camshaft and overhead-valve singles expanded considerably. With the retirement of Moto Guzzi, Mondial, and Gilera at the end of 1957, the 1958 Grand Prix season shaped up as a contest between Ducati and MV. Carlo Ubbiali and reigning champion Tarquinio Provini were formidable opponents, and Ducati prepared bikes for both the Italian and world championships. Farnè, Spaggiari, and Gandossi competed in the Italian championships with the team expanded to include Romolo Ferri, Francesco Villa, Luigi Taveri, Dave Chadwick, and Sammy Miller for the Grands Prix.

Desmodromic 125 Single and Twin

For 1958 the Desmo single featured new crankcases, strengthened at the rear to provide a support for the twin rear engine plates, and some included a six-speed gearbox. Development during the season resulted in the single-cylinder 125 Desmo, achieving 23 horsepower at 12,500 rpm. The 1958 Desmo frames were double-cradle, and they were very successful this year, Farnè winning the Junior Italian 125cc Championship, with Spaggiari taking out the Senior Italian 125cc Championship. In the World Championship, Alberto Gandossi finished second, with victories in Belgium and Sweden.

With the World Championship lost, Ducati then went to the final Grand Prix at Monza determined to at least gain some national kudos. Francesco Villa debuted the new 125 Desmo twin, and Desmo singles were provided to Gandossi, Taveri, Chadwick, and Spaggiari. In front of 100,000 enthusiastic spectators, Spaggiari led Ducatis home in the first five places, winning at an average speed of 155.828 kilometers an hour. For Taglioni and Ducati, it was a spectacular way to end the season and some consolation for losing the World Championship.

Dave Chadwick finished third in the 1958 125cc TT at the Isle of Man.

Romolo Ferri on his way to second place in the Isle of Man 125cc Grand Prix. Ferri's Desmo had a special lowered frame to suit his small stature. *Ducati*

Based on the earlier 175, the 125cc twin had a bore and stroke of 42.5x45mm, a 10.2:1 compression ratio, and a pair of special 23mm Dell'Orto carburetors. The power was 22.5 horsepower at 13,800 rpm. Unfortunately, the powerband was extremely narrow and the machine was difficult to ride. The frame was a twin-cradle type and the wheels 17-inch, but the 125 Desmo twin never handled as well as the single.

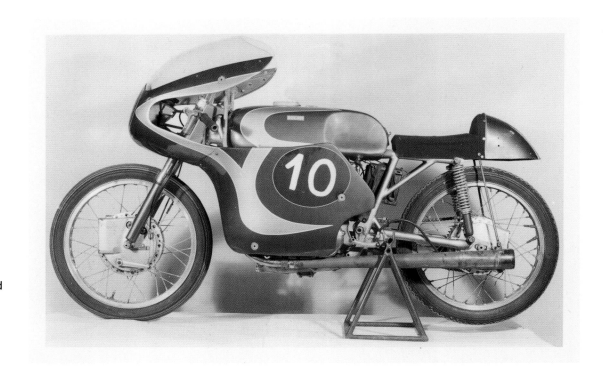

The 125cc Desmo twin debuted at the 1958 Nations Grand Prix at Monza, Francesco Villa finishing third behind the Desmo singles of Spaggiari and Gandossi. *Ducati*

125 Grand Prix Bialbero

By 1958, many of the developments of the Desmo also appeared on the Bialbero. These included new strengthened crankcases and in some cases twin spark plug ignition and the Desmos' double-cradle frame. As a privateer racer, the 125 Bialbero was extremely successful, and in England during 1958, Fron Purslow won a succession of races, this carrying on with teenage sensation Mike Hailwood after he purchased Purslow's machine.

The 125 Desmo twin was more complex than the single and not as reliable.

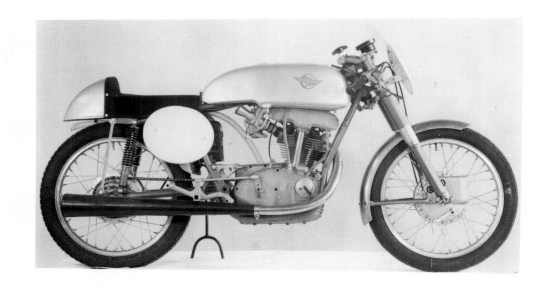

The 125 Grand Prix Bialbero received the Desmo's strengthened crankcases for 1958. *Ducati*

PRODUCTION OVERHEAD CAMSHAFT MODELS 1958

175 Sport, 175T

Updates to the 175 for 1958 included a new camshaft with revised timing, new primary drive gears, and a new carburetor. The front engine plates now included four mounts instead of three, with a new front downtube, the rear subframe incorporated with the main frame, and updated passenger footpeg mounts. Other than the new frame, the 175T continued largely unchanged during 1958. This updated frame was subsequently shared with the rest of the overhead camshaft range, including all 100, 125, and 200cc models until 1960.

175 SPORT, 175T 1958 Differing from 1957

Carburetion	UB22.5 BS 2
Engine and frame numbers	From DM175 76001 (175 Sport) From 02001 (175T)

175 Americano

Soon after securing American distribution, Berliner requested a specific American version, the 175 Americano. Initially this was an Americanized version of the 175 Super Sport, with a 125 Sport–style tank and high handlebars, but was joined by a more extravagant 175T-based model during 1958. Thus, there were two distinct series of Americano, each with a different engine and frame number sequence.

Available as an almost identical 200, the 175 Americano (Type 1) was an extremely attractive model. Powered by the uprated 175 Super Sport engine with larger carburetor and single muffler, the Americano also featured a crash bar and adjustable Marzocchi shock absorbers. As on the 175 Super Sport, the carburetor breathed through an open bell mouth, but the high handlebars restricted the claimed top speed to 85 miles per hour. Berliner offered this version to its 400-strong dealer network during 1959.

The second type was based on the 175T, with the deeper mud guards, engine in a lower state of tune, but still with the 13-liter 125-style tank. Styling additions extended to a "cowboy" handlebar, a stepped studded seat, and chrome tank panels. Some had matching leather saddlebags and air horns mounted on the crash bar. The 175T-based Americano also included the 175 Sport's dual muffler.

The 1958 175 Americano was extravagantly styled, with a studded seat and cowboy handlebars. Twin air horns were mounted on the large crash bar. *Ducati*

175 AMERICANO 1958
Differing from the 175 Sport and 175T

Carburetion	Dell'Orto SSI 25A (Type 1)
Power	18 horsepower at 8,000 rpm (Type 1)
Brakes (mm)	Drum 158 front, 136 rear
Dry weight (kg)	118 (Type 2)
Top speed (km/h)	137 (Type 1)
Colors	Blue and silver (Type 2)
Engine and frame numbers	DM175 from 77001–78000 (Approx.) Type 1 from 02501 Type 2

175 Motocross (Scrambler)

As with many other Italian motorcycle manufacturers during the 1950s, Ducati was involved in various off-road racing competitions. Following the creation of a 250cc Italian off-road championship in 1956, Ducati decided to adapt the 175 overhead camshaft single for off-road competition, and for 1958 the company released the 175 Motocross. As it was intended for competition, this was quite a serious attempt at creating a competitive off-road motorcycle and cross between a real dirt racer and trail bike.

But for the addition of an air cleaner, the 175 Motocross engine was the same as the first type 175 Americano. But compared to other 175s, the frame was significantly strengthened. In addition to full loops underneath the engine, the steering head was braced with additional gusseting, the swingarm was stronger, with a redesigned rear shock absorber mount and chain adjuster. The loop from the rear top shock mount was redesigned and an additional brace added under the seat. Other special features included a conical front brake, a longer front fork, and larger wheels.

125 Sport

Only a year after the introduction of the 175, the 125 Sport and similar 100 Sport were offered. The smaller capacity engines differed to the 175 in a number of details, notably the sump was cast smooth, without finning, and the bevel shaft outer tubes were also smooth cast and incorporated as one piece with the bearing housing.

Apart from a lower first and second gear, the clutch and gearbox was shared with the 175 Sport, but with a lower primary drive. The electrical system was the CEV of the 175 Sport, with the 175 Sport–style peaked headlight rim. The earliest 125 Sport shared a similar frame with the first series 175 Sport, with a three-bolt front engine mount and bolted-on subframe, but with narrower front engine mounts, a narrower and shorter rear subframe, and a shorter and less braced swingarm. Sometime during 1958, the frame became the later type with a three-bolt front engine mount.

The 175 Motocross included a number of special components, notably the frame, brakes, and suspension. *Ducati*

175 MOTOCROSS 1958
Differing from the 175 Sport

Carburetion	Dell'Orto SSI 25A
Power	14 horsepower at 8,000 rpm
Front suspension	Marzocchi fork
Brakes (mm)	Drum 180 front and rear
Wheels	21x2.25-inch, 19x2.25-inch
Tires	2.25x21, 3.00x19
Wheelbase (mm)	1,380
Dry weight (kg)	122
Colors	Copper and red
Engine Numbers	DM175 from 76001 (1958)

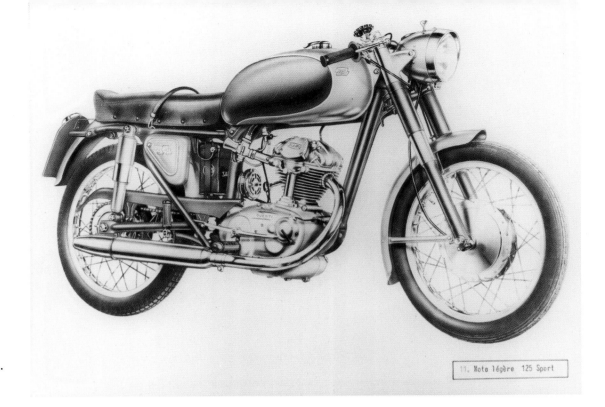

The early 1958 125 Sport had the first series frame and the larger brakes of the 175 Sport. *Ducati*

Initially the 180 and 160mm brakes were shared with the 175 Sport, but the aluminum wheel rims were 17-inch and soon the brakes were downsized to 158 and 136mm. Unlike the 175 Sport's sculptured fuel tank, the 125 Sport's 17-liter tank was styled along the lines of the F3.

125TS

Joining the 125 Sport for 1958 was the 125TS (Turismo Speciale). While ostensibly similar to the 125 Sport, the 125TS engine was in a lower state of tune, with a lower compression ratio and smaller carburetor. Inside the cylinder head were a number of differences to the 125 Sport, primarily the replacement of hairpin valve springs with a pair of coil springs.

125 SPORT 1958 Differing from the 175 Sport

Bore (mm)	55.2
Stroke (mm)	52
Displacement (cc)	124.443
Carburetion	Dell'Orto MB20B
Power	10 horsepower at 8,500 rpm
Front suspension	Ducati 30mm fork
Rear suspension	RIV or Marzocchi twin shock swingarm
Brakes (mm)	Drum 180 (158) front, 160 (136) rear
Wheels	17x2.25-inch front and rear
Tires	2.50x17, 2.75x17
Wheelbase (mm)	1,320
Dry weight (kg)	100.5
Top speed (km/h)	112
Colors	Gold and blue, gold and red
Engine and frame numbers	From 200001

125TS 1958 Differing from the 125 Sport

Compression ratio	7:1
Carburetion	Dell'Orto UA18BS
Power	8 horsepower
Brakes (mm)	Drum 158 front, 136 rear
Wheelbase (mm)	1,310
Dry weight (kg)	98.5
Top speed (km/h)	90
Colors	Blue and white, red and white
Engine and Frame Numbers	From DM125TS 650001

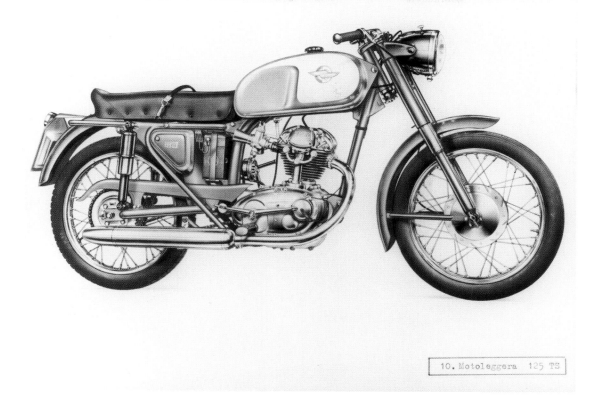

The 125TS was a detuned version of the 125 Sport and was the only model with coil valve springs. *Ducati*

The 125TS frame was shared with the 125 Sport, but the Marzocchi fork was a more basic design, with smaller diameter tubes and a pressed-steel outer cover incorporating the headlight ears. The black-painted Ceriani shock absorbers were nonadjustable and two versions were available, a Biposto and Monoposto.

100 Sport

Released alongside the 125 Sport during 1958 was the similar 100 Sport. But for a smaller piston and higher compression ratio, ostensibly the 100cc engine was identical to the 125 Sport but with a smaller carburetor.

Produced primarily for the Italian market, the 100 Sport was very similar to the 125. *John Goldman*

100 SPORT 1958
Differing from the 125 Sport

Bore (mm)	49
Displacement (cc)	98.058
Compression ratio	9:1
Carburetion	Dell'Orto MA18B
Power	8 horsepower at 8,000 rpm
Dry weight (kg)	100
Top speed (km/h)	105
Colors	Blue and silver
Engine and frame numbers	From 250001

TAGLIONI'S ARRIVAL

OVERHEAD VALVE SINGLES AND TWO-STROKES, 1958
125TV, 125S, 125T, 98TS, 98TL, 85 Sport, 85T, 65TS, 65TL, 65T

This year saw the end of the three Cucciolo-based 65s and the introduction of a downsized 98, the 85. The 98 continued as a new model, the 98TS (Turismo Speciale), and all 125s received a double-cradle frame, decals instead of tank badges, and bright red and white colors.

For 1958, the 85cc and 98cc overhead valve engines were redesigned, still with the external oil line on the right, but with a removable rocker cover that allowed for easier valve adjustment. This cover lacked the two-stroke-like finning of the earlier cylinder head and included reshaped engine covers with an enclosed flywheel on the left. The new 98TS featured a more mildly tuned version of the 98cc engine, with the 125 tubular steel frame, also with a double front downtube, with styling very similar to the 125T.

This year also saw the first Ducati two-stroke, the 48 Sport. The three-speed single-cylinder two-stroke engine hung from an open tubular steel frame. The cylinder was inclined 25 degrees, starting was by pedal, and the styling was similar to the 98 TL. But after only one year, the 48 Sport was discontinued, the two-stroke reappearing in 1962.

1959

Emphasis on developing the production range of overhead camshaft singles saw a considerably reduced racing effort for the 1959 season, with Desmo singles loaned to a selection of riders for various events, notably Mike Hailwood. Mike's wealthy father Stan arranged to take over the UK distribution of Ducati motorcycles and was able to obtain factory 125 Desmo singles and twins for the 1959 Grand Prix season. Mike was very successful on the single, with victories in the British 125 Championship, the Ulster 125cc Grand Prix, and third place in the 125cc World Championship.

85 TURISMO, 85 SPORT 1958–1961
Differing from the 98

Bore (mm)	**45.5**
Displacement (cc)	**84.55**
Compression ratio	**7.5:1**
Carburetion	**Dell'Orto ME15BS**
Power	**5 horsepower (5.5 horsepower, 85S)**
Gears	**Three-speed (Four-speed 85S)**
Tires	**2.50x17-inch**
Brakes (mm)	**116x25**
Weight (kg)	**80**
Top speed (km/h)	**76 (85S), 70 (85T)**
Colors	**85S silver and blue, blue frame**
	85T red with white, red frame
Engine and frame numbers	**From DM85S (85T) 500001**

For 1958, the 125T received a double-cradle frame. *Ducati*

98TS 1958–1963
Differing from the 98TL

Compression ratio	7:1
Carburetion	Dell'Orto ME16BS
Power	6 horsepower at 6,800 rpm
Gears	Four-speed
Tires	2.50x17-inch, 2.75x17-inch
Front suspension	Telescopic fork
Rear suspension	Swingarm
Brakes	123x25mm
Wheelbase (mm)	1,270
Weight (kg)	87
Top speed (km/h)	85
Colors	Red and white

TOP: The 85 Sport was more attractive than the 85T, but the performance was disappointing. *Ducati*

CENTER: Two 125 Sports were available in 1958, this is the overhead valve version with similar styling to the 85.

LEFT: With only a three-speed transmission, the 85T was a basic machine.

ABOVE: The updated engine featured an easily removable rocker cover and enclosed flywheel.

ABOVE RIGHT: New for 1958 was the 98TS. The engine and frame were similar to the 85's.

OPPOSITE TOP: Although it looked similar to the 125 Sport, the 125F3 was ostensibly a Gran Sport with enclosed valves. *Roy Kidney*

OPPOSITE BOTTOM: Light and reasonably powerful, the 175 was the most successful of the F3s. It also came with twin-scoop Amadoro brakes.

125 and 175F3

During 1958, the Formula 3 superseded the Marianna as a cataloged production racer but wasn't generally available until 1959. Offered initially as a 125 and 175, despite the overhead camshaft single entering regular production during 1957, the F3 was still largely based on the earlier Marianna. The 125F3 had Marianna sand-cast crankcases, and apart from a one-piece cylinder head with enclosed valve springs, it was ostensibly identical to the Marianna. As on the Marianna, the primary and bevel gears were straight cut.

Although patterned after that on the road versions, the single downtube frame was quite different. Lighter and lower, with a lower steering head, shorter swingarm, and shorter forks, they also varied between models, the 125 with the lowest steering head and shortest front fork. Both had 18-inch wheels, the 125 generally with Marianna brake and the 175 with twin-scoop Amadori, similar to the 125 Grand Prix. Despite their obvious race orientation, many F3s came with complete street equipment that included a headlight, muffler, taillight, number plate holder, and centerstand.

Although the F3s were genuine factory racing machines, they suffered because they were too expensive for most privateers and were penalized by the four-speed gearbox. Specifically designed for Formula 3 racing in Italy, the F3's first major success was at Monza in 1958 in the 175 F3 support race for the Nations Grand Prix. Here Franco Villa rode a factory-prepared 175F3 to win at an average speed of 142 kilometers an hour. In the wake of this success, Ducati sent Franco Farnè with a 175F3 to America early in 1959, Farnè winning six events in succession and proving unbeatable in the Lightweight classes at Daytona, Marlboro, and Laconia.

TAGLIONI'S ARRIVAL | 49

125, 175 FORMULA 3 1959–1962

Bore (mm)	55.25, 125 (62, 175)
Stroke (mm)	52 (57.8, 175)
Displacement (cc)	124.4 (174.5)
Compression ratio	9:1
Carburetion	Dell'Orto SSI 20C (125) SSI 22.5A (175)
Power	12 horsepower at 9,800 rpm (125), 19 horsepower at 9,700 rpm (175)
Gears	Four-speed
Wheels	18-inch
Front suspension	35mm telescopic fork

PRODUCTION OVERHEAD CAMSHAFT MODELS 1959

200 Élite

As the Berliners demanded a larger capacity motorcycle, Ducati responded by boring the 175 and creating the 200 for 1959. The first 200 was the Élite, and except for the larger capacity engine, deeper fenders, and steel wheel rims, this was very similar to the 1959 175 Sport. As these early 200 Élites were 175-based, they were denoted Type A, with "A" stamps on the cylinder and head. Carburetion on the first 500 engines was by a racing-style Dell'Orto SSI, while the exhaust system was a dual muffler type.

Although US examples received abbreviated fenders, for Europe these were deep and valanced on early Élites. First-edition 200 Élites also featured wider steel wheels with larger tires, and while the sculptured fuel tank was similar to the 175 Sport, it included chrome panels.

The first 200 Élite had deep fenders but retained the 175 Sport's clip-on handlebars.

200 ÉLITE 1959
Differing from the 175 Sport

Bore (mm)	67
Displacement (cc)	203.783
Compression ratio	8.5:1
Carburetion	Dell'Orto SS27A (until 150500) UBF24BS (from 150151)
Power	18 horsepower at 7,500 rpm
Front suspension	Ducati fork 30mm (31.5mm from 151501)
Rear suspension	Marzocchi twin shock swingarm
Wheels	18x2.50-inch
Tires	2.75x18, 3.25x18
Dry weight (kg)	106
Top speed (km/h)	140
Colors	Blue and aluminum
Engine and frame numbers	E150001–E150500

200 Super Sport

Released alongside the 200 Élite for 1959 was the 200 Super Sport. Like the Élite, early versions were blue and aluminum but with more sporting fenders and a single muffler. The 200SS always featured the slightly narrower wheels and smaller section tires, with aluminum wheel rims optional.

200TS, 200 Americano

Joining the 175 Americano and 175TS for 1959 was the 200 Americano and 200TS. The engine was shared with the 200SS and Élite, and both the TS and Americano came with the deeper Élite fenders and steel wheels with grooved block tread tires front and rear. But while the Americano included a higher handlebar, the 200TS shared the lower Élite handlebars. The 200TS Americano was ostensibly a variation on the earlier 175 Americano, continuing the same style but with fewer seat studs and no air horns on the crash bar.

200 Motocross (Scrambler)

Introduced at the same time as the 200 Super Sport and Americano, and also intended primarily for the American market, was the 200 Motocross. Similar to the earlier 175 Motocross, the 200 Motocross shared the 175's frame, with additional gussets and double cradle underneath the engine, conical brakes, and larger steel wheels. But while it should have been more popular, the 200 didn't sell as well as the 175.

The 1959 200 Super Sport was very similar to the Élite. Early examples were blue and silver. *Roy Kidney*

200 SUPER SPORT 1959
Differing from the 200 Élite

Front suspension	Marzocchi fork 32mm (Ducati 31.5mm from 151501)
Wheels	18x2.25-inch
Tires	2.50x18, 2.75x18

200TS, 200TS AMERICANO 1959
Differing from the 200 Élite

Power	17 horsepower at 7,500 rpm
Tires	3.00x18 front and rear block
Dry weight (kg)	118
Top speed (km/h)	120

ABOVE: Also new for 1959 was the 200TS, an amalgam of the Élite and Super Sport. *Ducati*

LEFT: The 200 Motocross was a development of the 175 and included many specific components.

200 MOTOCROSS 1959
Differing from the 200 Élite

Carburetion	Dell'Orto SS27A with F20A filter
Front suspension	Marzocchi 31.5mm fork
Wheels	21x2.25-inch and 19x2.50-inch
Tires	2.75x21 and 3.00x19
Wheelbase (mm)	1,380
Dry weight (kg)	124
Top speed (km/h)	107
Colors	Copper and red

175 SPORT, 175T, 175 MOTOCROSS 1959
Differing from 1958

Power	11 horsepower at 6,000 rpm (175T)
Brakes (mm)	Drum 158 front, 136 rear
Engine and frame numbers	DM175 from 77001 (175 Sport and Motocross) From 04001 (175T)

175TS 1959–1961
Differing from the 175T

Wheel	18x2.00-inch front
Wheelbase (mm)	1,310
Dry weight (kg)	108
Top speed (km/h)	110
Colors	Copper and red, blue and silver
Engine and frame numbers	DM175 from 04001

175 Sport, 175 Super Sport, 175T, 175TS, Americano, 175 Motocross

Beginning in 1959, a tuning kit was available for the 175 Sport that included a larger SSI 25A carburetor. In England the 175 Sport was called the 175 Silverstone and with the tuning kit was known as the 175 Silverstone Super. To further confuse the nomenclature, Berliner decided to market the 175 Sport as a Super Sport during 1959, with a single silencer and the tuning kit. The claimed power was 18 horsepower and the top speed 90 miles per hour.

For 1959, the 175T brakes were the smaller type of the 125 and 175TS, while options now included a single seat, high handlebar, and parcel carrier. The American 175T featured even higher handlebars, but the 175 Americano and Motocross were unchanged this year.

New for 1959 was the 175TS (*Turismo Speciale*), effectively another variation of the 175T. Large numbers were built, primarily for the domestic market, but the touring models were generally not popular elsewhere. Although ostensibly similar to the 175T, the 175TS included a more practical handlebar, lower seat, and smaller brakes (from the 125), also shared with the 175T this year.

125 Sport, 125TS, 100 Sport

Sometime during 1958, the 125 frame was changed to the updated four-bolt front engine mount type with one-piece rear subframe, and the brakes were downsized. For 1959, a decal celebrating Ducati's Montjuïc 24-hour race victories of 1957 and 1958 was included on the tank. In America, the 125 Sport was sold as the 125 Super Sport (with Sport decals), and in England the 125 Sport was called the 125 Monza. But as it was listed at $499 (in comparison to $599 for the 175), the 125 Sport didn't sell well. As the 125 remained popular in Italy, it continued for a few more years, but 1959 was the final year for the 100 Sport.

PUSHROD SINGLES 1959
125TV, 125T, 98TS, 98 Bronco, 85 Sport, 85T, 85 Bronco

With the end of the 65, the 85 assumed the role as the base model, the 85T becoming the 85 Bronco (with a smaller gas tank) for the United States. Also new for the United States this year was the 98 Bronco, sold in Europe as the 98 Cavallino.

125 SPORT, 125TS, 100 SPORT 1959
Differing from 1958

Carburetion	**Dell'Orto UB20BS (125 Sport)**
	Dell'Orto ME18BS (125TS)
	Dell'Orto UA18BS (100 Sport)
Power	**6.2 horsepower at 6,500 rpm (125TS)**
Brakes (mm)	**Drum 158 front, 136 rear**
Colors	**Red/copper (125 Sport)**

ABOVE: For 1959 the 125 Sport received a new frame and smaller brakes. *Roy Kidney*

RIGHT: The 1957 US 125 Sport was bronze and maroon, with tank badges rather than decals. *Roy Kidney*

TAGLIONI'S ARRIVAL

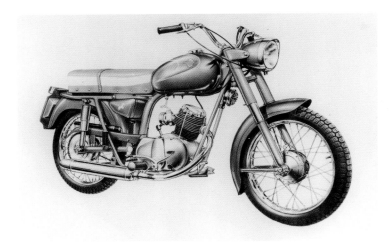

98 BRONCO 1959–1962
Differing from the 98TS

Tires	2.75x16-inch Bronco
Brakes	116x25mm Bronco
Colors	Purple and bronze

One of many models produced for Berliner was the 98 Bronco. A smaller fuel tank, wheels, and brakes set it apart from the 98TS. *Ducati*

This was very similar to the 98TS except for a 13-liter fuel tank and 16-inch wheels with smaller diameter brakes.

1960

At the end of 1959, Ducati ceased its official involvement in Grand Prix racing and Stan Hailwood managed to acquire a Barcone Desmo single. He also commissioned Taglioni to build Desmo 250 and 350 parallel twins based on the 125. The Barcone was the final development of the Desmo single, and along with larger six-speed crankcases, it featured a single downtube frame, splitting into double cradle underneath the engine. On English short circuits, Hailwood and the Barcone proved a formidable combination, Hailwood winning the 125cc British Championship.

Pressure from Berliner also had led to a 250cc version of the 175F3 in 1960. With a special 74mm cast-iron liner, the prototype 250F3 produced 32 horsepower at 9,000 rpm, and Francesco Villa was sent to contest a series of road races in the United States. So successful was this racing effort that it established Ducati in the United States as a manufacturer of reliable and competitive racing machines, also paving the way for the release of the production 250.

BELOW: While the triple camshaft cylinder head was similar to earlier versions, the Barcone (or Barge) crankcases were larger and stronger.

BELOW RIGHT: The Barcone 125 was the final version of the racing Desmo single.

ABOVE LEFT: Mike Hailwood with the 250 Desmo twin in 1960. *Ducati*

ABOVE: As it was basically an enlarged 125, the 250 twin's handling was flawed.

250 and 350 Desmo Twins

Early in 1960 Mike Hailwood received a 250cc parallel twin. The 250 shared the 55.25x52mm dimensions with 125 single but was otherwise a scaled-up version of the earlier 125 twin. With twin Dell'Orto 30mm carburetors, the power was 43 horsepower at 11,600 rpm, providing the six-speed 250 with a top speed of around 135 miles per hour. Unfortunately, the engine, weighing a claimed 112 grams, was too powerful and heavy for the flimsy double-cradle frame, and even with Norton forks and Girling shocks, the handling was deficient.

In an effort to improve the handling, John Surtees had Ken Sprayson build a new chassis with leading-link front suspension for the 250 and 350 Desmo twins.

TAGLIONI'S ARRIVAL | 55

Hailwood had some success on the 250 in England early in 1960 but was never satisfied with the handling, eventually commissioning an alternative frame.

Two 350cc twins were also built in 1960, one for Ken Kavanagh and another for Hailwood. The 64x54mm twin produced a claimed 48 horsepower at 11,000 rpm, but neither rider found it satisfactory. Early in 1961, Hailwood had a leading link Sprayson Reynolds frame built for the 350 and eventually John Surtees acquired all the 250 and 350 twins, converting them all with leading link frames. They weren't successful, the complex twins plagued with fractured crankcases, broken gears, and electrical and ignition problems before being retired during 1963.

PRODUCTION OVERHEAD CAMSHAFT MODELS 1960

With Berliner providing a healthy order book, production included 21,104 motorcycles this year; 14,403 of those were overhead camshaft singles. This would be the strongest production year for the overhead camshaft single.

200 Élite, 200 Super Sport, 200 Motocross, 200TS Americano

A number of updates were incorporated on the 200 for 1960, including a smaller carburetor and a new gearbox, clutch, and primary drive. Élite chassis updates included three-way adjustable shock absorbers, narrower steel wheel rims, and a smaller rear tire. The Élite fenders were now the more abbreviated and sporting 200SS type, accompanied by a shorter Aprilia taillight bracket. Ostensibly the Élite was now very much the same as the 200 Super Sport, set apart only by the dual muffler and specific fuel tank badges. The 200 Americano also continued unchanged this year, as did the 200 Motocross.

175 Sport, 175T, 175TS Americano, 175 Motocross

This year all 175s were basically sleeved-down 200s, sharing the same Type A crankcases and cylinder head castings, and inside the engine, the primary drive, clutch, and gearbox were extensively updated. The Marzocchi shock absorbers included three-way spring preload adjustment.

The 200 Super Sport for 1960 was in red and copper, still with chrome tank panels. It was very similar to the Élite this year.

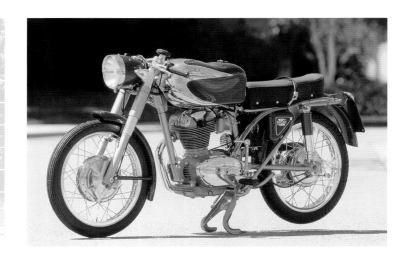

200 ÉLITE, SS, MOTOCROSS, AMERICANO 1960 Differing from 1959

Carburetion	Dell'Orto UBF24BS
Wheels	18x2.25
Rear Tire	3.00x18
Colors	Copper and red
Engine and frame numbers	From 150501
Production	1,486 (Élite) 843 (TS Americano) 607 (Motocross)

175 SPORT, T, TS AMERICANO MOTOCROSS 1960 Differing from 1959

Engine and frame numbers	DM175 from 78001 (175 Sport, Motocross) from 05001 (175T, TS)
Production	1,511 (175 Sport) 1,772 (175T USA) 1,162 (175 Americano) 1,466 (175 Motocross) 2,670 (175TS)

125 SPORT, 125TS

The 125 Sport also received a number of updates for 1960, including a new cylinder head, clutch, gearbox, and primary drive. Chassis updates included Radaelli steel wheel rims, and tank decals replaced the badges. The 125 Sport continued in this guise for several more years, primarily for the Italian market. The 125TS also included the engine updates but still retained coil valve springs. Two seat types were offered this year, striped or plain black.

PUSHROD SINGLES 1960
98TS, 98 Bronco (Cavallino), 85 Super Sport, 85 Sport, 85T, 85 Bronco

The range of pushrod singles now included an 85 Super Sport, with a higher compression ratio and larger carburetor, but only a few were manufactured. As it was competing with the more sophisticated overhead camshaft 125TS, the 125TV was dropped, while the more popular 85 and 98 pushrod singles continued unchanged.

The 125 Sport received attractive new colors for 1960 and remained popular in Italy until 1967.

125 SPORT, 125TS 1960 Differing from 1959

Front suspension	Ducati 31.5mm fork
Rear suspension	Ceriani
Colors	Gold/blue (125 Sport), Blue/gray (125TS)
Engine and frame numbers	From 203700 Approx. (125 Sport) From 650001 (125TS)
Production	671 (125 Sport) 890 (125TS) 1,325 (125TS black seat)

85 SUPER SPORT 1960
Differing from the 85 Sport

Compression ratio	9:1
Carburetion	Dell'Orto ME16BS
Power	6.5 horsepower at 7,500 rpm
Weight (kg)	79
Production	62 (85 SS)
Other pushrod production 1960	1,260 (85 Sport) 1,400 (85T) 748 (85 Bronco) 89 (98 Cavallino) 2,557 (98 Bronco) 585 (98TS)

TAGLIONI'S ARRIVAL | 57

CHAPTER 3
DUBIOUS PROLIFERATION
OVERHEAD CAMSHAFT, OVERHEAD-VALVE SINGLES, AND TWO-STROKES, 1961–1970

While the 250 motor was similar to the 175 and 200, the Type B engine castings were new. *Roy Kidney*

The 1961 250F3 was styled similarly to the 250 Diana. This year, the brakes were Amadoro, and although sold as a production racer, the 250F3 had a headlight and taillight.

As Ducati was already successfully campaigning factory 250cc racing machines, it was inevitable the overhead camshaft engine would grow to 250cc. The first production 250 was released during 1961, as the sporting Diana and touring Monza. This was typical of the evolutionary approach Ducati took at the time, and the overhead camshaft singles were secondary in production to the range of pushrod overhead valve singles.

At the same time, Ducati decided its future lay in higher volume cheap transportation, and during 1961 it was busy preparing a range of two-stroke singles. Ultimately, this was a flawed decision, and while Ducati managed to produce large numbers of these uninspiring models, it did little for the company's reputation and profitability.

1961

During 1961, Ducati homologated the Type B overhead camshaft 250, this with a new engine and frame sharing few parts with the previous Type A 200 and 175. The 200 and 175 continued as a Type A this year and also introduced was the catalog racing 250F3. The pushrod single range included the more sporting 125 Aurea, but overall production declined significantly from 1960, to 12,338.

250 F3

The 250F3, or Manxman, was a development of the 175F3, also sharing little with the production 250 Diana. The 250 cylinder head also came with bosses for desmodromic closing rocker spindles and a few were built as a Desmo. The 250F3 frame was similar to the 175F3, but with a slightly longer swingarm and a higher steering head, while the new 35mm Marzocchi fork had exposed stanchions and steel legs. The wheels were 19-inch, and the front brake for 1961 was a 200mm Amadori. Produced in very limited numbers, the heavier 250F3 was unable to replicate the 175F3's success.

250 FORMULA 3 1961–1962	
Bore (mm)	74
Stroke (mm)	57.8
Displacement (cc)	248.6
Compression ratio	8.5:1
Carburetion	Dell'Orto SSI 29A
Power	23 horsepower at 8,200 rpm
Gears	Four-speed
Wheels	19-inch

250 Diana (Daytona) and 250 Monza

The Type B overhead camshaft single was ostensibly a new model, the engine receiving new crankcases and other castings, and consequently sharing few parts with the earlier

250 DIANA (DAYTONA), MONZA 1961

Bore (mm)	74
Stroke (mm)	57.8
Displacement (cc)	248.589
Compression ratio	8:1
Carburetion	Dell'Orto UBF24BS
Power	19.5 horsepower at 7,550 rpm (16.4 horsepower at 7,200 rpm Monza)
Gears	Four-speed
Front suspension	Ducati 31.5mm fork
Rear suspension	Marzocchi twin shock swingarm
Brakes (mm)	Drum 180 front, 160 rear
Wheels	18x2.25-inch (18x2.5-inch Monza) front and rear
Tires	2.75x18, 3.00x18
Wheelbase (mm)	1,330
Dry weight (kg)	120 (125 Monza)
Top speed (km/h)	140 (135 Monza)
Colors	Blue frame, blue and silver (Diana) Blue/gold, Red/Silver, Black/Silver (Monza)
Engine and frame numbers	From DM250 80001

175-based Type A. For 1961 the two Type B 250s were the sporting Diana, or Daytona for the UK, and touring Monza. The frame was slightly redesigned for the Type B engine, with the swingarm bushes moved from the frame to the swingarm and the pivot locked in place by two pinch bolts. Diana and Monza frames in 1961 (along with Type B 200s and 125s of the same era) also included seat lugs on the crossbrace in front of the rear shock mount. The brakes were the usual seven-fin Grimeca, the front with cooling vents.

Setting the 250 Diana and Daytona apart from earlier sporting overhead camshaft singles was the new 17-liter steel fuel tank and steel side covers. These included intakes for the air filter and engine breather hoses, the left cover also containing a toolbox compartment with hinged cover. The similar 250 Monza was the touring model, retaining the smaller 13-liter 200TS fuel tank but including a higher handlebar, a side stand, and crash bar. Although little changed from the 200 and 175, the 250 Diana heralded a new era for Ducati. This was the first performance Ducati that was affordable and widely available, and with an optional tuning kit, it could be transformed into an effective club racer.

One of the classic narrow-case singles was the four-speed Diana. *Roy Kidney*

PRODUCTION 1961	1,560 (125TS) 1,000 (125 Sport) 1,283 (175TS) 115 (175 Motocross) 350 (200 Motocross) 250 (200TS Americano) 820 (200TS Americano black seat) 1,511 (200 Élite)

TYPE A SINGLES 1961
200 Élite, 200 Motocross, 200TS Americano, 175TS, 175 Motocross, 125 Sport, 125TS

The existing Type A overhead camshaft singles overlapped with the new Type B production during 1961, the range reduced. And apart from a small number of Americanos gaining a two-tone seat, the 200, 175, and 125 singles were unchanged from 1960.

PUSHROD SINGLES 1961
125 Aurea, 98TS, 98 Bronco (Cavallino), 85 Super Sport, 85 Sport, 85 Bronco

New for the pushrod single lineup this year was the sporting 125 Aurea, styled similarly to the higher specification 125 Sport. Similar to the earlier 125TV and retaining the double-cradle frame, complementing the 125 Sport-style gas tank was a sporty headlight, cigar-shaped muffler, and distinctive two-tone saddle.

The 250 Monza was the touring version of the Diana. *Ducati*

New for 1961 was the sporting 125 Aurea. Styled like the 125 Sport, the Aurea was more show than go. *Ducati*

125 AUREA 1961 Differing from the 125TV 1959	
Compression ratio	6.8:1
Carburetion	Dell'Orto ME18BS
Brakes (mm)	123x25
Wheelbase (mm)	1,285
Weight (kg)	90
Color	Metallic blue and gold, blue frame, blue seat
Engine and frame numbers	From DM125A 85001
Production	1,800
Other pushrod production 1961	40 (85 Sport) 22 (85 Super Sport) 1,200 (85 Bronco) 785 (98 Cavallino) 997 (98 Bronco) 605 (98TS)

1962

Although this year saw the consolidation of the new Type B overhead camshaft single, most resources were directed at implementing the new range of small capacity two-strokes. The first of these were three 48s, and overall production climbed to 24,443 this year. Designed to take advantage of new Italian highway regulations that didn't require licensing for 50cc motorcycles, the two-stroke exercise began strongly but was ultimately unsuccessful. But while the range of two-strokes was uninspiring, the 250cc overhead camshaft single evolved into the outstanding Scrambler and Diana Mark 3. In 1962, the 200 became the 250-based Type B, with two models continuing: the 200TS Americano and 200 Élite, with the 200GT joining them.

OVERHEAD CAMSHAFT SINGLES 1962

250 Scrambler (Motocross)

At the request of Berliner, a third 250 emerged during 1962, the 250 Scrambler. This replaced the 200 Motocross and also spawned the more sporting 250 Diana Mark 3, a superb production racer also specific to America. Unlike the 200 Motocross that featured a large number of unique parts, notably the brakes, suspension, and frame, the 250 Scrambler was basically a hot Monza.

Inside the Type B four-speed 250cc engine was a higher compression four-ring piston, hotter camshaft, and larger Dell'Orto SSI carburetor. The exhaust was by an open 36mm pipe. Unlike the 250 Diana (and other OHC singles), valve adjustment was now by shim rather than screw and locknut, while the ignition and 6-volt electrical system was from the 175F3, with a flywheel magneto, no battery or regulator, and extremely basic lighting. Berliner intended the Scrambler as a multipurpose model, terming it a "four-in-one" machine, and although fitted with knobby tires, the Scrambler could be adapted for street riding, road racing, short-track racing, or scrambles.

The 1962 250 Scrambler was based on the 250 Monza and very successful. The exhaust was open and the street equipment was barely legal. *Roy Kidney*

The 250 Scrambler engine was almost as highly tuned as that of the 250 Diana Mark 3. *Roy Kidney*

250 SCRAMBLER 1962
Differing from the 250 Diana/Daytona

Compression ratio	9.2:1
Carburetion	Dell'Orto SSI 27A
Power	27 horsepower at 8,000 rpm
Tires	3.00x19, 3.50x19
Wheelbase (mm)	1,350
Dry weight (kg)	109
Top speed (km/h)	135
Colors	Blue frame, blue and silver
Engine numbers	From DM250 80001–84500 Approx.
Production	417 (USA)

250F3, 250 Diana (Daytona), 250 Monza, 250 Diana Mark 3

For 1962 the few 250F3s available had a larger 230mm Oldani front brake, but the release of the more affordable 250 Diana Mark 3 virtually rendered the expensive F3 obsolete. The 250 Monza continued unchanged, and while the standard 250 Diana continued to be popular in Europe and the UK (where it was still the Daytona), Berliner requested a more sporting version for 1962. A racing machine out of the crate, the stark elemental style struck a chord with Americans, and the 250 Diana Mark 3 would come to define Ducati in the United States. The performance was also comparable to current 250cc twins, and with a top speed of 100 miles per hour, it was nearly as fast as most 500cc motorcycles.

Ostensibly the 250 Diana Mark 3 was an amalgam of the standard Diana and US-only Scrambler. The engine included the Scrambler's Dell'Orto SSI 27mm carburetor, but without an air filter, and a slightly higher compression ratio and hotter camshaft. As on the Scrambler, valve adjustment was also by shim, and the Diana Mark 3 was supplied with a Silentium muffler or black-painted reverse cone megaphone exhaust. As the electrical system was also shared

The 250 Diana Mark 3's small headlight and taillight were shared with the 250 Scrambler.

250 DIANA MARK 3 1962
Differing from the 250 Diana and SCR

Compression ratio	10:1
Power	30 horsepower at 8,300 rpm
Tires	2.50x18, 2.75x18
Wheelbase (mm)	1,320
Dry weight (kg)	110
Top speed (km/h)	177
Colors	Blue frame, blue and silver
Engine numbers	From DM250 80001–85851

The 250 Diana Mark 3 was one of the most competitive club racers available in the early 1960s but was still hampered by a four-speed transmission. Street equipment was extremely minimal. *Roy Kidney*

BELOW: While the standard Diana wasn't very popular in the United States, the 250 Daytona continued to be successful in England.

BELOW: The 1962 250 F3 had a large Oldani front brake but was still handicapped by a four-speed gearbox. All F3s had sand-cast engines and the dimensions varied between model.

BOTTOM: The only instrumentation was a central Veglia tachometer, and the clip-on handlebars allowed for a smooth-top triple clamp. *Roy Kidney.*

with the 175 F3 and 250 Scrambler, including a flywheel magneto and no battery, lighting was minimal. The 250 Diana Mark 3 provided exceptional performance, was outstandingly successful, and was the most significant Ducati yet to be sold in America.

200 Élite, 200GT, 200 Americano, 125 Sport, 125TS

As demand continued for the 200 Élite after the introduction of the 250, the 200 continued as a sleeved-down Type B from 1962. The general specifications were unchanged and the frame was also a Type B, noticeable for the stronger swingarm pivot and pinch bolts. Another 200 was available for 1962, the 200GT, styled like the new 250 Diana with the Diana-type fuel tank, but with the earlier D-type decals. As it offered only moderate performance, the 200GT didn't initially prove popular, but it soon caught on, as insurance was cheaper for 200cc machines. Another variation of the 200 for 1962 was the 200 TS/Americano, now ostensibly a 250 Monza with a 200 Type B engine, but production only lasted for 1962. The 125 Sport and TS also continued unchanged this year.

BELOW: For 1962, the 200 Élite was a downsized 250 instead of an enlarged 175.

200 ÉLITE 1962 Differing from the Type A

Compression ratio	7.8:1
Dry weight (kg)	111
Top speed (km/h)	140
Engine and frame numbers	From DM200E 155000 Approx.
Production	1,400

200 GT 1962
Differing from the 200 Élite

Carburetion	UBF24BS
Dry weight (kg)	115
Top speed (km/h)	135
Color	Red and white
Production	280

200 TS AMERICANO 1962
Differing from the Type A

Compression ratio	7.8:1
Color	Red
Production	1,211

OTHER OHC PRODUCTION 1962
770 (125TS)
1,100 (125 Sport)

PUSHROD PRODUCTION 1962
850 (85 Bronco)
655 (98 Cavallino)
1,050 (98 Bronco)
900 (98TS)

PUSHROD SINGLES 1962
98TS, 98 Bronco (Cavallino), 85 Bronco

Production for the 125 Aurea went into hiatus this year, and to accommodate the new two-strokes, the range of pushrod singles rationalized.

TWO-STROKES 1962
48 Brisk, 48 Piuma, 48 Sport

New two-strokes for 1962 included the three-speed Piuma moped.

Ducati introduced three two-strokes for 1962, the 48cc engine first appearing in the short-lived 48 Sport of 1958. The single-speed 48 Brisk (a misnomer if there ever was one) was the basic model, and produced primarily for the Italian market, its small rack was designed to appeal to students. While the 48 Brisk had only a solo seat and pedal starting, the telescopic front fork and twin shock absorber rear swingarm was quite sophisticated for this type of machine. Alongside the Brisk was the similar Piuma (or Puma for the UK) with a three-speed unit gearbox and a hand gear change incorporated in the throttle grip. Apart from the transmission, and the Piuma's more angular styling, the Brisk and Piuma were virtually identical in specification.

Alongside the Brisk and Piuma for 1962 was the 48 Sport. Unlike the basic Brisk and Piuma, the 48 Sport was a mini motorcycle, with a tubular steel duplex cradle frame and clip-on handlebars. Starting was by kick or pedals, but setting the 48 Sport apart from the mopeds was the higher compression

The 48 Sport was styled to replicate the overhead camshaft Diana. *Ducati*

48 BRISK, 48 PIUMA, 48 SPORT	
Engine	Single-cylinder piston-port 2-stroke, air-cooled
Bore (mm)	38
Stroke (mm)	42
Displacement (cc)	47.633
Compression ratio	6.3:1 (9.5:1 Sport)
Carburetion	Dell'Orto T4 12D1 (UA 15S Sport)
Power	1.34 horsepower (1.5 Piuma, 4.2 Sport)
Gears	Single-speed (Three-speed Piuma & Sport)
Front suspension	Telescopic fork
Rear suspension	Twin shock swingarm
Brakes (mm)	Drum 90 (105 Sport)
Tires	2.00x18 (2.25x18 Piuma, 2.25x19 Sport)
Dry weight (kg)	45 Brisk, 47 Piuma, 49 Sport
Top speed (km/h)	40 Brisk, 50 Piuma, 80 Sport
Colors	Red, red and bronze (Sport)
Engine numbers	From DM48 010001, 300001 (Sport)
Frame numbers	From DM48P 01001 (Piuma)
	From DM48E 10001 (Sport)
Production	6,000 (48 Brisk) 5,500 (48 Piuma) 4,000 (48 Sport)

engine with a downdraft Dell'Orto carburetor and polished alloy bell mouth. The three-speed gearshift was still hand operated.

1963

The proliferation of two-stroke models this year saw production climb to a staggering 37,034 motorcycles, and as there was no longer an official racing program, it was the V-four Apollo that created most interest this year. The development of the overhead camshaft singles stalled, and the range of pushrod singles curtailed. But with Berliner calling the shots and management placing so much faith into the two-stroke range, this was an uneasy time at Borgo Panigale. So many two-strokes were produced that unsold stock became a serious concern and motorcycle production wouldn't exceed the 1963 figure until the year 2000.

Apollo

Apollo, the Greek god of the arts, is representative of order, harmony, civilization, and prophesy, and there was no more prophetic a motorcycle than the extraordinary 10-horsepower 1,260cc V-four Ducati Apollo. That this motorcycle was designed and built in Italy, to American requirements, by a company with little experience of motorcycles over 250cc was even more amazing. At that time the Apollo reeked of extreme excess, yet fifty years later, its size and specification can be considered mainstream. The Apollo was truly a motorcycle ahead of its time.

Joe Berliner wanted to break Harley-Davidson's stranglehold on the police motorcycle market in the United States. Official police department specifications required motorcycles to have engines 1,200cc or larger, a wheelbase of at least 60 inches, and tires of a 5.00x16-inch section. He convinced Montano and Taglioni to consider such a project, agreeing to finance the cost of the prototype and assist in the cost of tooling for production. Berliner wanted a

APOLLO

Engine	90-degree V-four 4-stroke, air-cooled
Bore (mm)	84.5
Stroke (mm)	56
Displacement (cc)	1,257
Compression ratio	10:1 (8:1)
Valve type	Overhead valve, pushrod
Carburetion	Dell'Orto SSI 32 (SSI 24)
Power	100-horsepower (80-horsepower), later 67 horsepower
Gears	Five-speed
Final drive	Duplex chain
Front suspension	38mm Ceriani fork
Rear suspension	Twin shock absorber swingarm
Tires	5.00x16-inch Pirelli
Wheelbase (mm)	1,550
Dry weight (kg)	240

TOP: Decades ahead of its time, the Apollo was too large and powerful for the tires of 1964.

ABOVE: Looking at the Apollo engine, it is easy to see the inspiration for the later V-twins.

lightweight, 1,200cc V-four motorcycle that would outperform the Harley in every respect. Ease of maintenance was also a priority, and this called for pushrod and rocker overhead valves, with the gear-driven single camshaft positioned between the V, just like an American V-8. The initial Apollo produced a then-shattering 100 horsepower, the engine incorporated as a stressed member in a massive frame. Unfortunately, no tires were available in 1963 that could cope with sustained 120-mile-per-hour operation, and although the power was subsequently restricted, Ducati and Berliner simply didn't have the resources to implement production. Fortunately, Taglioni wasn't discouraged by the Apollo's failure and was able to use many of its features as an inspiration for the 750 V-twin seven years later.

FOUR-STROKE SINGLES 1963
200GT, 250 Diana, 250 Diana Mark 3, 250 Monza, 250 Scrambler, 125 Aurea, 98TS

With 1964 scheduled for more updates, the overhead camshaft lineup was reduced to only five models for 1963, mostly unchanged. The only 200 was the 200GT, with no 175s or even 125s this year. While the existing 250s continued as before, the 250 Monza received a new seat, with curved aluminum beading, and was two-tone on some examples. Only two pushrod singles were produced this year, the 125 Aurea making a brief return with the 98TS continuing for one more year.

TWO-STROKES 1963
48/1 Brisk, 48 Piuma, 48 Piuma Sport, 48SL, 48 Cacciatore, 48 Sport (50 Falcon), 80 Setter, 80 Sport, 80 Falcon

Now representing the bulk of production, a variety of 48s was built this year, the 48 Brisk evolving into the 48/1 Brisk, now styled similarly to the Piuma, with a Piuma Sport also available. The engine specification for all 48s included a slightly higher compression ratio and more power, with some examples of the 48 Piuma Sport having the higher output 48 Sport engine. Also available for 1963 was the Piuma Export, with an integrated headlight, dual seat, larger brakes, and fully enclosed drivechain. US examples featured the higher specification 48 Sport engine.

OHC AND PUSHROD PRODUCTION 1963
911 (200GT)
500 (250 Diana)
383 (250 Monza)
250 (250 Scrambler USA)
1,000 (125 Aurea)
500 (98TS)

Apart from a new seat, the 250 Monza was little changed for 1963. *Ducati*

Other new models included the 48 Sport Export, the sculptured gas tank mirroring that of the 200 Élite (a few with the new-for-1964 specification fan-cooled motor), the 48SL, off-road 48 Cacciatore, and the 50 Falcon (a 48 Sport) for the United States. The trio of 80s included the 80 Setter, 80 Sport, and 80 Falcon. Styled similarly to the domestic market 48 Sport, the 80 Setter included a single downtube frame (using the engine as a stressed member) and deeper fenders. A higher performance 80 Sport was also available this year, with clip-on handlebars and a more sporting gas tank, along with the semi-off-road 80 Falcon. The 80s proved unpopular in America and England and were only available in Italy after 1963.

ABOVE: The 80 Setter had a single downtube frame utilizing the engine as a stressed member. *Ducati*

ABOVE RIGHT: The 48 Sport Export had an Élite-style fuel tank and shock absorbers with exposed springs. *Ducati*

48/1 BRISK, PIUMA, PIUMA SPORT 48 SPORT, SL, CACCIATORE
Differing from 1962

Compression ratio	7:1
Power	1.5 horsepower
Engine and frame numbers	From DM 700001 (48SL) From DM48SL 0001 (Frame)
Production	8,500 (48/1 Brisk) 3,500 (48 Piuma pedal) 3,812 (48 Piuma kick) 2,300 (48 Piuma Export pedal) 49 (48 Sport pedal) 6,488 (48 Sport kick) 5,500 (48 Sport Export kick) 561 (48 Sport Export fan)

80 SETTER, 80 SPORT, 80 FALCON
Differing from the 48

Bore (mm)	47
Stroke (mm)	46
Displacement (cc)	79.807
Compression ratio	7.1:1
Carburetion	Dell'Orto ME 15BS
Power	4.25 horsepower (5.5 HP)
Gears	Three-speed
Brakes (mm)	Drum 118
Tires	2.25x18, 2.50x17
Dry weight (kg)	62
Top speed (km/h)	75
Colors	Blue and silver
Production	580 (80 Setter) 1,500 (80 Sport) 700 (80 Falcon)

Although retaining a pressed-steel frame, the 48 Piuma Sport was an attractive offering.

1964

After a two-stroke-driven manufacturing explosion during 1963, production returned to a more realistic 6,788 for 1964. Heading the lineup were new five-speed overhead camshaft singles, but while Ducati was preparing for their introduction, many of the existing four-speed models continued as before. The first five-speed models were the 250GT and 250 Mark 3, followed by the Mach 1, and with the four-speed catalog racing 250F3 now outclassed, Ducati released the five-speed Mach 1S production racer. An interim four-speed Scrambler was also introduced this year along with another dubious attempt at market penetration with the Brio scooter.

FIVE-SPEED OVERHEAD CAMSHAFT SINGLES 1964

250GT

The 250GT was the replacement for the four-speed 200GT, with a similar chassis but a new five-speed engine. Along with the new gearbox, the electrical system was improved, the flywheel alternator now rated at 60 watts, and the engine was in a milder state of tune. Unique features included the high swan-neck style handlebars individually mounted underneath the top triple clamp. As the power was extremely modest, the low performance 250GT only lasted for 1964.

The 250GT was the first five-speed overhead camshaft single but was extremely underpowered. *Ducati*

250 Mark 3

Along with the introduction of the five-speed 250GT, a five-speed 250 Mark 3 replaced the four-speed 250 Diana Mark 3 for the USA. No longer officially termed a Diana, these now featured high handlebars mounted on the upper triple clamp instead of racing clip-ons. The taillight now included a toggle

DUBIOUS PROLIFERATION | 71

250GT 1964
Differing from the 250 Monza

Power	18.4 horsepower at 7,200 rpm
Gears	Five-speed
Wheelbase (mm)	1,320
Colors	Red and white, black and silver
Engine numbers (on left crankcase)	DM250 85852–87421
Frame numbers	DM250 85852–87421 US bikes didn't have a frame number but a foil attached to the head stem with the same number as the engine.
Production	722

250 MARK 3 1964
Differing from 1963

Engine numbers	DM250 85852–92171

switch to ground the electrical system if a taillight bulb blew, this allowing the magneto ignition to function. Apart from the five-speed transmission, the engine was unchanged from 1963, retaining the flywheel magneto alternator and black-painted reverse cone megaphone.

250 Mach 1

The success of the four-speed 250 Diana Mark 3 in America prompted Ducati to create the five-speed Mach 1 in 1964. Unlike the Mark 3, the Mach 1 was intended as a sporting street bike to be sold worldwide, and as it needed to be street legal, it included the 250GT's battery electrical and ignition system. With its claimed top speed of

ABOVE RIGHT: For 1964 the five-speed 250 Mark 3 had high handlebars and a new taillight. Standard equipment included a tachometer and racing plate. *Ducati*

With a battery-powered electrical system, the 250 Mach 1 was a street-legal version of the US-only 250 Mark 3.

250 MACH 1 1964
Differing from the 250 Mark 3

Carburetion	Dell'Orto SSI 29D
Power	27.6 horsepower at 8,500 rpm
Wheelbase (mm)	1,350
Dry weight (kg)	116
Top speed (km/h)	170
Colors	Red and silver
Engine numbers	DM250M1 00001–00370 Approx.
Frame numbers	DM250 00001–00370 Approx.
Production	362

106 miles per hour and unique red frame, the Mach 1 would become the most revered of the narrow-case production singles.

Powering the Mach 1 was an uprated five-speed 250GT engine, with larger valves, a longer duration, higher lift camshaft, and a new Borgo three-ring forged piston. The inlet port was enlarged to match the larger Dell'Orto carburetor, and the running gear was ostensibly the same as the Mark 3 but for a red frame and rear-set footpegs and controls. The gas tank was the smaller and slimmer 200GT type, indented underneath to provide clearance for the larger carburetor, and most Mach 1s had clip-on handlebars with a flat-top triple clamp. A bizarre option was a high US touring-style handlebar.

Claimed to be the fastest 250 available, the Mach 1 has become one of the more legendary production Ducatis.

Mach 1S #00007 in as-raced condition. Originally raced by Frank Calder in Connecticut in 1964, this bike has a nonstandard seat, but the original lighting system remains. *Phillip Andersen*

250 Mach 1S

Only available to factory, distributor, and dealer racing teams, a small number of Mach 1Ss was produced during 1964. These looked similar to the regular Mach 1 but carried 250M1S stamps on the engine and a specific number sequence beginning at 00001 and finishing around 00030 Approx. Inside the engine were a number of special factory components, notably an F3 crankshaft with dual-rib con rod and larger valves. Ignition was by magneto, like the Mark 3. While the front fork and single downtube frame were shared with the Mach 1, the front brake was generally a 200mm Oldani. It was rumored five Mach 1Ss were imported by Berliner.

DUBIOUS PROLIFERATION | 73

During the 1960s, Berliner's publicity was very effective. Here *Cycle* magazine publisher Floyd Clymer shakes the hand of 250 Monza-mounted 1964 Indianapolis 500 race winner A. J. Foyt Jr.

250 SCRAMBLER 1964
Differing from 1963

Power	30 horsepower at 8,300 rpm
Engine numbers	From DM250 84500 Approx.–85851
Production	107
Other Four-Speed Overhead Camshaft Single Production 1964	498 (250 Diana) 67 (250 Monza USA) 610 (200 Élite) 766 (200GT) 452 (125 Sport) 670 (125TS)

FOUR-SPEED OVERHEAD CAMSHAFT SINGLES 1964
250 Scrambler, 250 Monza, 250 Diana/Daytona, 200 Élite, 200GT, 125 Sport, 125TS

While the four-speed 250 Daytona and Monza continued unchanged, the 250 Scrambler was restyled for 1964. Along with a special Giuliari Scrambler saddle was a longer and lower Diana-style fuel tank, and more practical rubber covered folding footpegs. The 200 Élite was offered as a Grand Sports or Super Sports in the UK and, along with the 200GT, a special import by Ghost motorcycles in New York, while after a one-year hiatus, both the 125 Sport and 125TS made a return, unchanged from 1962.

PUSHROD SINGLES 1964
125 Bronco

The only pushrod single produced this year was the US-only 125 Bronco. Other than a smaller 13-liter gas tank and 16-inch wheels shod with knobby tires, this was very similar to the Aurea. Unfortunately, the combination of very modest power in a moderately heavy chassis made for an extremely slow motorcycle, even by the standards of the 1960s.

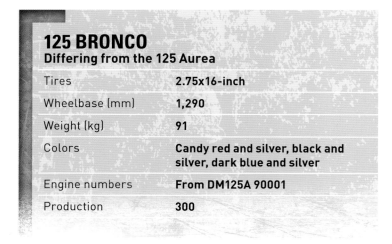

One of the mainstays of the lineup in the United States during the 1960s was the 125 Bronco, but it wasn't very popular. *Ducati*

125 BRONCO
Differing from the 125 Aurea

Tires	2.75x16-inch
Wheelbase (mm)	1,290
Weight (kg)	91
Colors	Candy red and silver, black and silver, dark blue and silver
Engine numbers	From DM125A 90001
Production	300

ABOVE LEFT: The 50 Falcon received fan cooling for 1964, the exhaust exiting on the left. Only a few were made. *Ducati*

ABOVE: Also fan cooled, the 90 Cadet, or Falcon, had the exhaust on the right. *Ducati*

TWO-STROKES 1964
50 Falcon, 80 Setter, 80 Sport, 90 Falcon, 90 Mountaineer, 90 Cacciatore, 48 Brio

As so many 48cc variants were produced in 1963, they continued to be available for several more years unchanged. The existing 80 Setter and Sport continued as before, but only in Europe, while the other two-strokes received fan cooling for 1964. An alloy plate bolted to the flywheel magneto drove the fan, and a metal shroud enclosed the cylinder and cylinder head. The carburetor was no longer downdraft, and the frame was again the dual-cradle type.

American demand for larger capacity models led to the 90 Cacciatore (Mountaineer) and 90 Cadet (Falcon) late in 1963, as 1964 models. The 87cc fan-cooled engine was ostensibly identical to the 48cc unit but with a higher output alternator and larger carburetor. The left engine cover incorporated a large orifice to allow hot air to escape, the exhaust exiting on the right rather than the left. Both the 90 Cadet and Mountaineer had a hand gear change, dual-cradle frame, and solo seat with rack.

The final dubious models for 1964 were the Brio scooter and Fattorino Carrier. Intent on creating as many two-stroke variants as possible, the 48cc fan-cooled engine was drafted into the Brio scooter and 3R Fattorino (errand boy) Carrier late in 1963. Harking back to the ill-fated Cruiser of 1952, the Brio was a genuine scooter in the Vespa and Lambretta mold, while the Carrier was a commercial three-wheeler available in several guises. These uninspiring machines were created primarily for the Italian market, but small numbers were exported. Unfortunately for Ducati, by the time the Brio was released, the scooter boom had passed, and Brios weren't even particularly popular in Italy.

1965

For 1965 all the overhead camshaft singles received a five-speed gearbox, and at Berliner's request a 350 joined the ten-model lineup. Motorcycle production increased slightly this year, to 7,440. With the United States continuing as Ducati's primary market, in February the factory sent a team

90 CADET, 90 CACCIATORE (MOUNTAINEER) 1964

Engine	Single-cylinder piston-port two-stroke, fan-cooled
Bore (mm)	49
Stroke (mm)	46
Displacement (cc)	86.744
Compression ratio	8.5–9:1
Carburetion	Dell'Orto UA 18S
Power	6 horsepower
Gears	Three-speed
Brakes (mm)	Drum 118
Tires	2.25x18, 2.50x18 (2.50x16, 3.25/3.50x16, Cacciatore)
Wheelbase (mm)	1,160
Dry weight (kg)	66 (68 Cacciatore)
Top speed (km/h)	90
Two-Stroke Production 1964	22 (50 Falcon) 480 (80 Setter) 789 (80 Sport) 50 (90 Falcon) 71 (90 Mountaineer) 822 (90 Cacciatore)

48 BRIO 1964

Engine	Single-cylinder piston-port two-stroke, fan-cooled
Bore (mm)	38
Stroke (mm)	42
Displacement (cc)	47.633
Compression ratio	7:1
Carburetion	Dell'Orto SHA 14/12
Power	1.5 horsepower
Gears	Three-speed
Front suspension	Swinging shackle fork
Rear suspension	Swingarm with rubber shock absorbers
Brakes (mm)	Drum 105
Tires	2.75x9
Dry weight (kg)	63.5
Top speed (km/h)	50

ABOVE: Unable to learn from the disastrous Cruiser experience, Ducati released another scooter, the 48 Brio for 1964. *Ducati*

BELOW: Franco Farnè testing the 125cc four-cylinder racer early in 1965. *Ducati*

to Sebring, Franco Farnè winning the 350cc class on a new 350SC. The recently released 350 was subsequently titled the Sebring.

Also this year Taglioni resurrected an earlier four-cylinder 125cc racing design. Created at the request of the Spanish Mototrans subsidiary, the 125 four had an eight-speed gearbox, four valves per cylinder, and double overhead camshafts driven by a central chain of spur gears. The power from the 34.5x34mm four was initially 23 horsepower at 14,000 rpm, but although the claimed weight was a low 85 kilograms, after initial testing in early 1965, the 125cc four disappeared during 1966. During the 1970s and 1980s, it was on display in the Leningrad Museum but eventually returned to Italy.

250 and 350SC

For 1965, the 250SC and 350SC replaced the Mach 1S as the catalog racer. Like the earlier F3, the SC was a hand-built racing machine that shared little with the production 250, the engine including sand-cast crankcases designed to accommodate the double-cradle frame. The sand-cast cylinder head was shared with the 250F3, while the close-ratio five-speed gearbox was unique to the SC, with wider gears and longer shafts. Each SC came with its individual wiring diagram and optional magneto ignition.

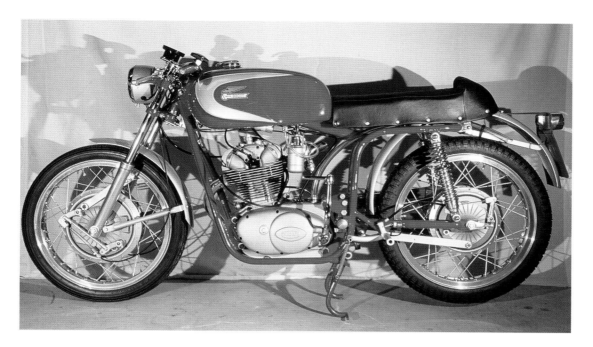

The 250SC featured a double-cradle frame and for 1965 a Mach 1–style fuel tank.

Apart from the double-cradle frame with wider engine mounts that was specific to the SC, there was also a special alloy Marzocchi fork with stepped fork tubes. The steel fuel tank was patterned after the Mach 1, and as with the F3, there was full street equipment, including a horn. Only around forty 250SCs and ten 350SCs were built for the 1965 season.

OVERHEAD CAMSHAFT SINGLES 1965
350 Sebring, 350 Scrambler, 250GT, 250 Monza, 250 Scrambler, 250 Mark 3, 250 Mach 1, 200 Élite, 200GT, 125 Sport

Apart from the four-speed 200 Élite, 200GT, 125 Sport, and the new 160 Monza Junior, all overhead camshaft singles received a five-speed gearbox this year. In the wake of criticism that it was underpowered, the 250GT engine was uprated to Monza specification, with a new camshaft and larger intake valve. Except for the new five-speed engine, the 250 Monza continued much as before, with a smaller 13-liter round gas tank, as did the 250 Scrambler. A number of similar 350 Scramblers were also built this year.

As Ducati already had the higher performing Mach 1 in Europe, for 1965 the US 250 Mark 3 received the Mach 1 engine. The engine now included the Mach 1's larger valves and SSI carburetor. Retained from the earlier Mark 3 was the flywheel magneto ignition, the electrical system still running without a battery. Also differing from the 1964 Mark 3 was the addition of side covers, and the adoption of the Mach 1's rear-set footpegs and controls. The Mark 3's red and silver colors and black frame were unchanged from 1964, as were

250 AND 350SC 1965

Bore (mm)	74 (250), 76 (350)
Stroke (mm)	57.8 (250), 75 (350)
Displacement (cc)	248.6 (340)
Compression ratio	10.2:1 (10:1)
Carburetion	Dell'Orto SSI 30 (SSI 32B, 350)
Brakes	230mm Grimeca
Wheels	19-inch
Wheelbase (mm)	1,320
Weight (kg)	115 (250) 117 (350)
Top speed (km/h)	190 (250) 200 (350)
Engine Numbers	DM250SC 1–40 Approx. DM350SC 1–10 Approx.

250GT, 250 MONZA, 250 & 350 SCRAMBLER 1965
Differing from 1964

Engine and frame numbers	From DM250 87422
Production	710 (250GT) 152 (250 Monza) 163 (USA 250 Scrambler) 121 (USA 350 Scrambler)
Production (200 and 125 1965)	381 (200 Élite) 569 (200GT) 513 (125 Sport)

ABOVE: A 1965 Berliner advertisement showing a 250GT (not a Monza as stated) in front of the Unisphere at New York World's Fair. This is a prototype model; the production examples had the newer Eagle tank decals.

ABOVE RIGHT: The 250 Mark 3 received a Mach 1 engine for 1965. New were the rear-set controls and Mach 1–style side covers. *Ducati*

BELOW: Dominated by the Dell'Orto 29mm SSI carburetor, the Mach 1 engine continued unchanged for 1965. Unlike the Mark 3, Mach 1s didn't have a tachometer drive from the top bevel gear cover.

250 MARK 3 AND MACH 1 1965
Differing from 1964

Carburetion	Dell'Orto SSI 29D (Mark 3)
Wheelbase (mm)	1,350 (Mark 3)
Dry weight (kg)	112 (Mark 3)
Engine numbers	From DM250M1 00370 Approx.
Production	338 (Mach 1)

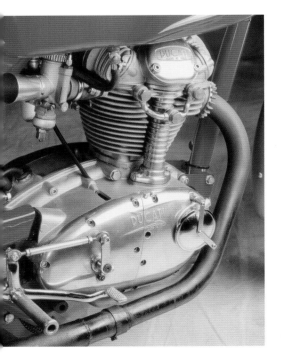

the standard high handlebars, Veglia tachometer, racing plate, and megaphone exhaust. Other than a new seat, the Mach 1 continued unchanged.

350 Sebring

At Berliner's request, the five-speed Type B engine grew to 350cc for 1965, but as Taglioni was worried about the reliability of the lower end, the 350 Sebring was conservatively tuned. The mild camshaft and small-valve cylinder head was shared with the 250 Monza and GT, but the 350 had a three-ring Borgo piston and higher primary drive ratio. The basic chassis was also that of the 250 Monza, and two versions were offered for 1965. Except for colors and side cover decals, the US edition was identical to the 250 Monza while the second 1965 edition was styled like the 250 Diana/Daytona, with a larger 17-liter tank and bench seat.

160 Monza Junior

Another model produced at Berliner's request was the 160 Monza Junior. Based on the 125, the 160 Monza Junior engine was externally identified by a lack of finning on the sump or rocker covers. The electrical system included a flywheel magneto for the ignition, similar to

350 SEBRING 1965
Differing from the 250 Monza

Bore (mm)	76
Stroke (mm)	75
Displacement (cc)	340.237
Compression ratio	8.5:1
Carburetion	Dell'Orto UBF24BS
Power	32 horsepower at 8,500 rpm
Wheelbase (mm)	1,330
Dry weight (kg)	123
Top speed (km/h)	125
Engine numbers	From DM350 01001
Production	1,286

the 250 Scrambler but with a small battery to power the lights. Contributing to the reduced overall dimensions were the 16-inch steel wheels with smaller brakes, and two versions were offered for 1965. The first was styled like the 250 Monza (with a round tank), and the second with an angular tank and fenders but retaining the earlier bullet-shaped Aprilia headlight.

Tuned conservatively and styled like the 250 Monza, the 1965 350 Sebring provided disappointing performance. *Ducati*

The first 160 Monza Junior was styled similarly to the 1965 250 Monza.

160 MONZA JUNIOR 1965
Differing from the 125 Sport

Bore (mm)	61
Stroke (mm)	52
Displacement (cc)	156
Compression ratio	8.2:1
Carburetion	Dell'Orto UB22BS
Power	16 horsepower at 8,000 rpm
Wheels	16x2.25-inch, 16x1.85-inch
Tires	2.75x16, 3.25x16
Wheelbase (mm)	1,330
Dry weight (kg)	106 (108 from second series 1965)
Top speed (km/h)	102
Colors	Black/silver, red/gold
Engine numbers	From DM160 18001
Production	710

ABOVE: The 100 Mountaineer was very similar to the 90, but for the American market, the number 100 had a better ring to it. *Ducati*

ABOVE RIGHT: The 100 Brio had smaller diameter wheels and a larger seat than the 48 Brio.

PUSHROD SINGLES AND TWO-STROKES 1965
125 Bronco, 80 Setter, 80 Sport, 100 Cadet, 100 Mountaineer, 90 Cadet (Falcon), 90 Mountaineer, 100 Brio

While the 125 Bronco, 90 Cadet (Falcon), 90 Mountaineer, 80 Setter, and 80 Sport continued unchanged, a slightly larger 100 Cadet and Mountaineer replaced the 90s during 1965. A 100cc Brio also joined the 48 this year, now with a dual seat and smaller diameter wheels. For the 100 Cadet and 100 Mountaineer, the fan-cooled engine was bored to 94cc, but the power was unchanged. The 100 Cadet now featured a dual seat.

1966

Ducati offered mild updates to the range of overhead camshaft singles, with the two-stroke lineup continuing largely unchanged. Motorcycle production slipped to 5,913, but more serious problems lay ahead when the American sales boom collapsed in the fall of 1966.

100 CADET, 100 MOUNTAINEER 1965
Differing from the 90 1964

Bore (mm)	51
Displacement (cc)	94
Compression ratio	10:1
Production 1965	455 (100 Cadet)
	369 (100 Mountaineer)
125 Bronco and Other Two-Stroke Production 1965	911 (125 Bronco)
	80 (90 Cadet)
	80 (90 Falcon)
	110 (90 Mountaineer)
	220 (80 Setter)
	200 (80 Sport)

100 BRIO 1965 Differing from the 48 Brio

Bore (mm)	51
Stroke (mm)	46
Displacement (cc)	94
Compression ratio	8.5:1
Carburetion	Dell'Orto SHB 18/16
Power	7 horsepower
Tires	2.45x8, 3.50x8
Dry weight (kg)	80
Top speed (km/h)	76

250 and 350 SC

Still available to special order through distributors, the production racing 250 and 350SC were little changed for 1966. The 250SC now received an Amal carburetor (the 350 retained the Dell'Orto), and both versions came with Oldani brakes front and rear. Also new was a fiberglass tank and humpback solo seat, but the basic dimensions were unchanged. The SCs were undoubtedly extremely beautiful creations, but they were too heavy and the double-cradle frame offered little advantage over the standard single downtube type. They weren't spectacularly successful but paved the way for the next racer, the Desmo, debuted by Farnè at Modena in March 1966. This first desmodromic 250 was a modified 250SC, with the double-cradle frame, but when it became a catalog racer for 1967, the Desmo was more closely related to the regular production machines.

The 1966 250SC had a new fuel tank, Oldani brakes, and an Amal carburetor. *Ducati*

250 AND 350SC 1966
Differing from 1965

Carburetion	Amal 1 3/16-inch 389 monobloc (250)
Brakes	200mm Oldani

OVERHEAD CAMSHAFT SINGLES 1966

350 Sebring, 350 Scrambler, 250 Monza, 250GT, 250 Scrambler, 250 Mark 3, 250 Mach 1, 200 Élite, 160 Monza Junior, 125 Sport

For 1966 the 250 and 350 overhead camshaft singles received new Grimeca three-rib brakes, the front with an additional vent on the right, this brake subsequently lasting until 1972. The 250 and 350 touring models now featured the dubious angular styling introduced on the 160 Monza Junior during 1965, and this year would see the final development of the narrow crankcase range.

250 Mark 3 and 250 Mach 1

The 250 Mark 3 was little changed for 1966, but because of criticism of the kick-start arrangement with the rear-set footpegs, the Diana and Monza's rocking gearshift lever and forward mount pegs returned. Apart from the new brakes, the Mach 1, never as popular as anticipated, was unchanged for its final year.

250 MARK 3, MACH 1 1966
Differing from 1965

Engine numbers (250 Mark 3)	DM250M3 or M1 01500–01800 Approx. Also from DM250M3 92172–105000 (1966–1967)
Production	138 (Mach 1)

RIGHT: For 1966, the 250 Mark 3 reverted to the previous rocking gearshift.

DUBIOUS PROLIFERATION | **81**

The 1966 brochure still depicts the Mark 3 (on the left) with rear-set footpegs.

The front brake from 1966 also included a right-side dummy air scoop.

350 Sebring, 250 Monza, 250GT, 200 Élite, 160 Monza Junior, 125 Sport

The 250 and 350cc touring models were restyled along the lines of the second 1965-series 160 Monza Junior, an angular headlight joining the angular fuel tank and fenders. The bench seat was thicker, and the 250 Monza and GT were ostensibly identical. Although the new styling wasn't universally accepted, these unexciting models sold surprisingly well, particularly the 160 Monza Junior. The 125 Sport continued unchanged, still primarily for the Italian market, while the four-speed 200 Élite was available ostensibly for the UK, where it was sold as a 200 Super Sport. Unlike the five-speed singles, the 200 Élite retained the earlier front brake (without dummy air scoop).

250 and 350 Scrambler

Ducati endeavored to widen the Scrambler's appeal for 1966, making it more suitable for street use by equipping it with the 160 Monza Junior's electrical system. This included the 160's bullet-style headlight and taillight, a new flywheel magneto, and a small battery to power the stoplight and parking lights. Other updates for 1966 included modified suspension, an 18-inch rear wheel, and more road-oriented universal tires. In America the range of optional equipment included a speedometer and muffler, with a range of cables and sprockets standard with each machine. Somehow these updates lessened the Scrambler's appeal, and it didn't sell as well this year.

350 SEBRING, 250 MONZA, 250GT, 160 MONZA JUNIOR, 200 ÉLITE, 125 SPORT 1966

Production	1,166 (350 Sebring)
	760 (250 Monza)
	126 (200 Élite)
	406 (160 Monza Junior)
	377 (125 Sport)

ABOVE LEFT: New angular styling was featured on the touring models for 1966. This is the most popular, the 160 Monza Junior.

250 & 350 SCRAMBLER 1966
Differing from 1965

Rear wheel	18x3.00
Rear tire	4.00x18
Dry weight (kg)	109
Engine and frame numbers	DM250 92172–105000
	DM350 05006–05105
Production	203 (250 Scrambler)
	145 (350 Scrambler)

PUSHROD SINGLES AND TWO-STROKES 1966
125 Bronco, 100 Cadet, 100 Mountaineer, 90 Cadet (Falcon), 90 Mountaineer, 50SL, 100/25 Brio

Both the fan-cooled 100 Cadet and 100 Mountaineer received a four-speed gearbox with right side foot shift, via the usual rocking pedal, but were otherwise unchanged. The Brio 100 evolved into the Brio 100/25, and although they had been out of production for several years, the Brisk, Piuma Sport, 50SL, and Brio 48 were still available in Europe this year. The 48 Piuma Sport became the 50 Piuma Sport but was otherwise unchanged.

Other new models for 1966 were the 50SL and 50SL/2. The 50SL was another bland rendition of the earlier 48SL, but with a new four-speed air-cooled engine and the angular styling (fenders and headlight) that was prevalent that year. The 50SL/2 had its own distinctive sculptured gas tank but was a touring model with high handlebars and a high exhaust pipe. These heralded a new range of sporting 50cc two-strokes for 1967.

LEFT: Updates to the 250 Scrambler for 1966 were intended to make it more street-oriented.

50SL, SL2 1966

Engine	Single-cylinder piston-port two-stroke, air-cooled
Bore (mm)	38.8
Stroke (mm)	42
Displacement (cc)	49.660
Compression ratio	9.5:1
Carburetion	Dell'Orto SHA 14/12
Power	4.2 horsepower
Gears	Four-speed
Tires	2.25x19
Dry weight (kg)	58
Top speed (km/h)	80
Production	10

100 CADET, 100 MOUNTAINEER 1966
Differing from 1965

Gears	Four-speed
Production	577 (100 Cadet)
	418 (100 Mountaineer)
125 Bronco and Other Two-Stroke Production	725 (125 Bronco)
	150 (90 Cadet)
	110 (90 Falcon)
	202 (90 Mountaineer)

100/25 BRIO 1966 Differing from the 100 Brio

Compression ratio	10:1

For 1966, the 100 Cadet received a four-speed transmission with a foot shift. *Ducati*

The air-cooled 50SL heralded a new range of small-capacity two-strokes. *Ducati*

1967

Despite the collapse of the US market, Ducati persevered with the development of the new wide crankcase single Ducati. For some reason the company also believed survival lay in the expansion of the two-stroke lineup and the introduction of a new overhead-valve single. All the new models saw production rising slightly to 6,242, but production didn't translate into sales and nearly 3,500 bikes destined for America eventually ended up in England, where they were available until 1971.

250 and 350SCD (Sport Corsa Desmo)

With the 250F3 cylinder head already set up for desmodromic valve gear, it was obvious the SC would receive a desmo head at some stage. Six 250s and one 350 were entered at Daytona in 1967, one for Farnè in the Expert Lightweight race, but the desmodromic Ducatis failed the AMA technical inspection as the valve system was considered "a change in basic design." In the end, Ducati received more publicity through not racing against the Yamahas.

These SCDs were quite different than the earlier SCs and were a precursor to the new production-wide crankcase singles that appeared at the end of 1967. Rather than special sand-cast crankcases, to save weight and cost, the SCD featured regular die-cast crankcases with a standard ratio five-speed gearbox. The SCD also had a single downtube frame, constructed of chrome-molybdenum without a loop over the swingarm pivot. The wheels were now 18-inch front and rear, with special shorter 35mm Marzocchi forks with alloy fork legs and a steel lower triple clamp. The brakes were Oldani. Only around twenty-five SCDs were produced. In the United States, Reno Leoni developed some into successful racers, while in Italy Roberto Gallina and Gilberto Parlotti achieved some respectable results in the early season Italian Riviera street races.

250 AND 350 SCD 1967–1968

Bore (mm)	74 (250), 76 (350)
Stroke (mm)	57.8 (250), 75 (350)
Displacement (cc)	248.6 (340)
Valve actuation	Single desmodromic overhead camshaft driven by a shaft and bevel gears
Power	35 horsepower at 11,500 rpm (250 SCD) 41 horsepower at 10,500 rpm (350 SCD)
Gears	Five-speed
Front suspension	Marzocchi 35mm fork
Rear suspension	Twin shock absorber swingarm
Brakes	Oldani
Wheels	18-inch
Engine numbers	From SCD01

The 250 and 350SCD were ostensibly factory racers, the frame and engine design serving as a prototype for the production wide-case single.

DUBIOUS PROLIFERATION | 85

LEFT: Berliner's advertising during 1967 continued to emphasize the relationship with car racing.

ABOVE: By 1967, the 250 Mark 3 had moved away from the minimalist street racer of the earliest examples. The headlight was now shared with the Scrambler.

BELOW: The Mark 3 replaced the Mach 1 for 1967 and included the 250 Monza's ignition system.

OVERHEAD CAMSHAFT SINGLES 1967
350 Sebring, 350 Scrambler, 250 Monza, 250 Scrambler, 250 Mark 3, 160 Monza Junior, 125 Sport

With the introduction of the wide-case single imminent, the existing overhead camshaft singles continued much as before. The 200 was discontinued, but the four-speed 160 and 125 continued unchanged this year. The 250 Mark 3 continued to evolve from the original elemental sportster of 1962. As it replaced the Mach 1 in Europe, the 250 Mark 3 received the Monza flywheel, stator plate, and regulator. It also shared the Scrambler's headlight.

PUSHROD SINGLES 1967
125 Cadet/4, 125 Cadet/4 Lusso, 125 Cadet/4 Scrambler

The final pushrod overhead-valve Ducati single was the 125 Cadet/4, introduced in 1967 and lasting until 1970. Many cycle parts were shared with the two-stroke Cadet, but the engine was still based on the earlier overhead valve unit. The bore and stroke were new, as was the cylinder head design. The spark plug was moved to the right, and the two overhead valves were set parallel, rather than opposed as before. The flywheel alternator electrical system was shared with the 160 Monza Junior while the double-cradle tubular steel frame was similar to the 125 Bronco. Alongside the Cadet/4 was an almost identical Lusso (luxury), with an improved front fork, sporting bullet-style headlight, and larger tires. The Scrambler featured fork gaiters, a high-rise exhaust, and an engine protection plate.

Redesigned slightly, the 125 OHV engine lived on in 1967 as the 125 Cadet/4. The angular fenders and headlight were styling features of the period. *Ducati*

350 SEBRING, 350 AND 250 SCRAMBLER, 250 MONZA, 250 MARK 3, 160 MONZA JUNIOR, 125 SPORT 1967

Production	807 (350 Sebring)
	320 (350 Scrambler)
	210 (USA 250 Scrambler)
	192 (Mark 3)
	419 (250 Monza)
	922 (160 Monza Junior)
	263 (125 Sport)

125 CADET/4 (SCRAMBLER, LUSSO) 1967

Engine	Single-cylinder four-stroke air-cooled
Bore (mm)	53
Stroke (mm)	55
Displacement (cc)	121.3
Compression ratio	8.4:1
Valve type	Overhead-valve, pushrod
Carburetion	Dell'Orto ME18BS
Gears	Four-speed
Tires	2.25x18, 2.50x18
	2.50x18 2.75x18 (Lusso)
	2.75x18, 3.25–3.50x16 (Scrambler)
Brakes (mm)	Drum 118
Front suspension	Marzocchi fork
Rear suspension	Swingarm
Wheelbase (mm)	1,160
Weight (kg)	72 (75, Scrambler)
Top speed (km/h)	95 (80, Scrambler)
Colors	Red/white, black/white
Engine and frame numbers	From DM125 50001 (engine)
	From DM125C4/01001 (frame)
Production 1967	300 (125 Cadet/4 1966)
	554 (125 Cadet/4 1967)
	68 (125 Cadet/4 USA)
	160 (125 Cadet/4 Lusso)
	73 (125 Cadet/4 Lusso USA)
	170 (125 Cadet/4 Scrambler)
	160 (125 Cadet/4 Motocross USA)

Alongside the 125 Cadet/4 was the Cadet/4 Scrambler, but this was hardly a true off-road machine.

TWO-STROKES 1967
100 Cadet, 100 Mountaineer, 90 Falcon, 90 Mountaineer, 50SL, 50SL/1, 50SL/2, Brisk 50/1, 48SL, 48 Cacciatore, 100/25 Brio, Rolly

A sporting 50SL/1 joined the 50SL and 50SL/2 for 1967. The long gas tank featured twin filler caps and the front fork had exposed springs. The cylinder head was new, the chrome-plated cylinder including cross-ports and the compression ratio higher. The carburetor breathed through an alloy bell mouth and had the option of a low or high exhaust pipe with heat shield. It may have only displaced 50cc, but with its low handlebar and abrupt saddle, the 50SL/1 was one of the most purposeful-looking Ducatis of the era. The fan-cooled 48cc engine also continued this year in the 48SL, now with the sporting chassis of the 50SL/1.

For 1967, the two-stroke Cadet and Mountaineer received their fourth engine development in as many years, although the earlier fan-cooled versions were sold in the United States through 1967. The revised engine had a new cylinder head and chrome-plated cylinder, and

It may have only displaced 50cc, but the long gas tank and twin filler caps accentuated the 50SL's sporting nature.

it was no longer fan-cooled. Another bore increase boosted displacement, and along with a larger carburetor, the power increased slightly. Apart from the new engine, both the 100 Cadet and Mountaineer were unchanged.

The unremarkable Brisk became the Brisk 50/1 for 1967. Except for styling updates that included a square gas tank, angular front fender, and smaller version of the 160 Monza square-bodied headlight, this was identical to the 48/1. Also this year the 48 Piuma became the 50 Piuma, now with the angular gas tank and headlight of the similar Brisk that year. Styling touches extended to white-walled tires. Joining the 50 Piuma was the Rolly, with the Brisk's single-speed engine in a steel chassis but with no rear suspension. The front fork was a rudimentary hydraulic telescopic and the headlight the later Piuma angular type.

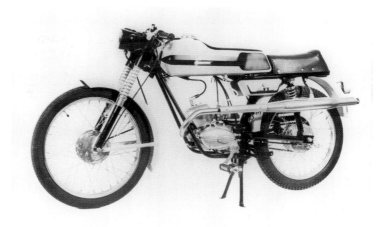

50SL/1 1967
Differing from the 50SL

Compression ratio	11:1
Carburetion	Dell'Orto UA018S
Power	6 horsepower
Dry weight (kg)	61
Top speed (km/h)	80
Engine numbers	From DM48V4 323001
Frame numbers	From DM48V4 323001 From DM48SL4 52001
Production	61

ABOVE: One of the more interesting two-strokes of the era was the 1967 48SL. While still retaining the earlier fan-cooled engine, the gas tank and exposed fork springs lent an air of purpose. *Ducati*

LEFT: The 1967 100 Cadet featured yet another engine with slightly more displacement, and it was air-cooled.

BOTTOM LEFT: Sharing the new engine was the 100 Mountaineer, but otherwise this was very similar to the previous model.

BOTTOM RIGHT: By 1967, the Brisk has evolved into the 50/1, with an angular gas tank and headlight. *Ducati*

100 CADET 100 MOUNTAINEER 1967
Differing from 1966

Engine	Single-cylinder piston-port two-stroke, air-cooled
Bore (mm)	52
Displacement (cc)	97.69
Compression ratio	11.2:1
Carburetion	Dell'Orto UBF24 BS
Power	7.2 horsepower at 6,250 rpm
Colors	Red frame, red and silver (Mountaineer)
Two-Stroke Production 1967	800 (100 Cadet) 611 (100 Mountaineer) 123 (90 Falcon) 164 (USA 90 Mountaineer)

ROLLY 1967–1968
Differing from the Brisk 50/1

Rear suspension	Rigid
Tires	2.00x18
Dry weight (kg)	42
Top speed (km/h)	50

ABOVE: Also new for 1967 was the uninspiring Rolly moped. *Ducati*

1968

Despite a difficult sales environment, Ducati released two significant models for 1968: the wide-crankcase overhead camshaft single and the Mark 3 Desmo. The wide-case single incorporated many improvements initiated with the racing SCD, while with the Mark 3 Desmo, Fabio Taglioni was able to realize his dream—the first production motorcycle with desmodromic valve operation. The new singles were introduced alongside the existing range of pushrod and two-stroke singles, overlapping with considerable unsold stock from the previous year, production increasing to 8,754.

OVERHEAD CAMSHAFT SINGLES (WIDE-CASE) 1968
250 and 350 Scrambler

Although ostensibly the same design as the previous narrow-case, the updated engine incorporated some updates aimed at improving reliability. The most notable new feature was the wider rear crankcase mount, but also significant was a strengthened kick-start, always problematic on the five-speed narrow-case. At the same time the entire engine structure was enlarged, and although reliability increased substantially, so did the weight. Accompanying the new engine was a larger and heavier frame. Many enthusiasts lamented the change to the wide-case, some racers still preferring the lighter and more responsive narrow-case, but ultimately the wide-case evolved into some of Ducati's finest models.

Despite the serious sales slump in America during 1967, the first of the new wide-case models was the Scrambler, aimed squarely at the US market. This was produced as a 250 and 350, and it was initially styled like the previous narrow-case. The 250 was in the same moderate state of tune as its narrow-case predecessor while the 350 featured a hotter cam,

250 & 350 SCRAMBLER 1968
Differing from 1967

Bore (mm)	74 (76)
Stroke (mm)	57.8 (75)
Displacement (cc)	248.589 (340.2)
Compression ratio	9:1 (9.5:1)
Carburetion	Dell'Orto SSI 27D (250)
	Dell'Orto SSI 29D (350)
Wheelbase (mm)	1,380
Dry weight (kg)	132 (133)
Colors	Red, white, black
Engine numbers (on right crankcase)	DM250 104501–106000 Approx.
	DM350 05006–05495
Production 1968	320 (350 Scrambler 1967)
	997 (USA 250 Scrambler)
	865 (USA 350 Scrambler)

ABOVE LEFT: The first wide-case 250 Scrambler was styled similarly to the previous narrow-case. Also like the previous version, the exhaust was without a muffler. *Ducati*

LEFT: The second series 1968 250 Scrambler included a new fuel tank. The electrical system included a small battery and horn. *Ducati*

a higher compression ratio, and a larger carburetor. The new Marzocchi suspension included fork gaiters, but the wheels and brakes were unchanged.

Shortly after its introduction, both the 250 and 350 Scrambler were updated, with a more powerful 70-watt electrical system with a regulator and larger battery. The 11-liter fuel tank was also reshaped and included chrome panels. At some stage during 1968, the rear frame section was modified to include a loop connecting both rear downtubes.

250 Monza, 350 Sebring, and 160 Monza Junior

For 1968, the 250 Monza and 350 Sebring were also offered as a wide-case, the 250 Monza in two versions, one for the United States and another for Europe. European versions continued the angular style of the previous narrow-case, as did the 350 Sebring. As the engine specification of both these models was unchanged from the previous narrow-case and the weight increased, the performance was sluggish and both these models were unsuccessful. Although a large number of 160 Monza Juniors were amongst the bikes reexported to the UK, a few hundred Monza Juniors were built this year, unchanged from 1967.

The 350 Sebring was also briefly produced as a wide-case. The early frame here doesn't have a strengthening loop on either side near the swingarm pivot. *Ducati*

DUBIOUS PROLIFERATION | 91

250 MONZA, 350 SEBRING, 160 MONZA JUNIOR 1968–1969
Differing from 1967

Engine numbers (on right crankcase)	From DM250 104501 From DM350 05006
Production 1968	225 (350 Sebring) 887 (250 Monza) 401 (160 Monza Junior)

250 and 350 Mark 3

A more significant release for 1968 was the Mark 3, and similar Mark 3 Desmo, intended as sporting successors to the previous Mark 3. The engines were basically shared with the respective Scrambler, but the 250 was in a higher state of tune. This included the early Mach 1 and Mark 3 camshaft and larger valves, a higher compression piston, and a larger carburetor. The chrome steel wheels and brakes were shared with the previous Mark 3, as was the Marzocchi fork and shock absorbers with exposed springs.

For 1968, the Leopoldo Tartarini—designed Mark 3 fuel tank had twin filler caps, as on the 1967 on the two-stroke 50SL/1. However, the 13-liter breadbox-style fuel tank didn't meet with universal acclaim, and many Mark 3s were supplied with Scrambler tanks. Mark 3s also had silver-painted tank panels and silver-painted fenders.

One of the first of the new wide-case singles, the 1968 350 Mark 3.

250 & 350 MARK 3 1968-1970
Differing from the Scrambler 1968

Compression ratio	10:1
Carburetion	Dell'Orto SSI 29D
Wheels	18x2.50-inch front and rear
Tires	2.75x18, 3.00x18
Front suspension	Marzocchi 31.5mm fork
Wheelbase (mm)	1,360
Dry weight (kg)	127 (128)
Top speed (km/h)	143 (250) 150 (350)
Colors	Red/silver, red or black frame
Engine numbers	DM250 105001-108620 DM350 05105-11052
Production 1968	710 (350 Mark 3) 826 (250 Mark 3)

ABOVE: Only the 1968 Mark 3 and Desmo had the twin filler gas tank.

BELOW: The 1968 350 Mark 3 Desmo looked similar to the standard Mark 3. Only the early examples had a red frame.

250 and 350 Mark 3D

The 1968 production Desmo was ostensibly a Mark 3 with a desmodromic cylinder head, this patterned after the SCD, with a single four-lobe camshaft but retaining the hairpin closing valve springs, these the lighter 160 Monza Junior type. In all other respects, the Desmo engine was identical to the Mark 3, with the same valve sizes and Dell'Orto carburetor. Although the chassis, including the skinny Marzocchi front fork and twin filler fuel tank, was also shared with the Mark 3, the Mark 3 Desmo's tank featured chrome panels, small "D" decals on the side covers, and chrome-plated fenders and headlight shell.

As on the regular Mark 3, Scrambler fuel tanks were fitted to many 1968 250 and 350 Mark 3 Desmos. Several high-performance options were available on request, including a high handlebar, hotter camshaft, megaphone exhaust, alternative carburetor jets, and full racing fairing. When fitted with the racing kit, the 350 Mark 3 Desmo was claimed to be Ducati's fastest production model yet, with a top speed of 112 miles per hour. Despite its failings, in particular the weight and marginal suspension, the Mark 3 Desmo was one of Ducati's finer offerings of the 1960s.

250 & 350 MARK 3 DESMO 1968-1970
Differing from the Mark 3

Power	34 horsepower at 7,500 rpm (crankshaft 350) 20.5 horsepower (rear wheel 350) 36 horsepower (crankshaft 350 race kit)
Top speed (km/h)	150, 160 with megaphone (250) 165, 180 with megaphone (350)
Colors	Red, cherry red, red or black frame
Production 1968	309 (350 Mark 3D) 902 (250 Mark 3D)

PUSHROD SINGLES AND TWO-STROKES 1968
125 Cadet/4, 125 Cadet/4 Lusso, 125 Cadet/4 Scrambler, 100 Cadet, 100 Mountaineer, 50SL/1A, 50SL/2A, 100/25 Brio, Rolly

Although the 125 Cadet/4 and Cadet/4 Lusso received new side covers, the four-model lineup of pushrod singles continued largely unchanged for 1968. For 1968 the air-cooled 50cc two-stroke evolved into the 50SL1/A and 50SL/2A, the 50SL1/A with a new single filler gas tank and low exhaust. The SL2/A was a touring version, with high handlebars, shrouded fork springs, and dual seat. The 100 Cadet, 100 Mountaineer, 100/25 Brio, and Rolly continued unchanged, and the 90cc two-strokes were discontinued.

1969

After more than a decade of successive financial losses, mainly due to a dependence on the failing US market underwritten each year by the Italian government, Ducati was given a lease of life when it was absorbed into the EFIM (*Ente Finanzaria per gli Industrie Metalmeccaniche*) Group. New directors Arnaldo Milvio and Fredmano Spairani replaced Montano, and Ducati embarked on an ambitious developmental program. In the meantime, Ducati released its

PUSHROD SINGLE AND TWO-STROKE PRODUCTION 1968
621 (125 Cadet/4)
265 (125 Cadet/4 Lusso)
239 (125 Cadet/4 Scrambler)
456 (100 Cadet)
320 (100 Mountaineer)

RIGHT: The 125 Cadet/4 received new side covers for 1968 but was otherwise unchanged. *Ducati*

BELOW LEFT: The 50 SL1A replaced the 50 SL1 for 1968. *Ducati*

BELOW RIGHT: Joining the 50 SL1/A was the touring 50 SL 2/A. *Ducati*

largest capacity production model yet, the 450 Scrambler, still aimed at the US market. This evolved into the 450 Mark 3 and Desmo, and overall production increased to 11,295 during 1969.

450 Scrambler (Jupiter), 450 Mark 3, and 450 Mark 3 Desmo

When the wide-case overhead camshaft single was introduced in 1968, it was designed to allow a safe increase in capacity beyond the 340cc of the narrow-case, and during 1969, the single grew to 436cc. Although the American market was still in a slump, the first 450 was a US-intended Scrambler. Designed to compete directly with the likes of the British 441 BSA Victor, the 450 Scrambler was sold as the Jupiter in America, and although it wasn't initially very successful, it eventually gained a cult following in Europe. Incorporating many developments from the racing program, the first 450 Scrambler prototypes were running by July 1968, with production versions available from January 1969.

A new type of Dell'Orto square-slide VHB carburetor replaced the racing-type SSI, and while the cylinder head was basically unchanged, inside the new crankcases was a longer con rod and cylinder. The 450 frame included additional steel gusseting on the top tube and rear downtubes, but the brakes, wheels, and suspension were shared with the 250 and 350 Scrambler.

Alongside the 450 Scrambler for 1969, Ducati offered a Mark 3 and Mark 3 Desmo, these ostensibly the same as the respective 250 and 350 versions, but with the 450 engine and braced frame. The 450 Mark 3 engine was identical in specification to the 450 Scrambler, with the 450 Desmo similar to the 250 and 350 Desmo. This year the 13-liter coffin-style tank now included only a single filler cap, and like the 250 and 350, many 450 Mark 3s and Desmos were produced with a Scrambler tank. Most criticism of the new 450 centered on vibration and performance. The 29mm carburetor was considered very small for the engine's capacity and the 450 Desmo wasn't as fast as the earlier 250 Mach 1 and Mark 3.

Also new for 1969 was the Mark 3 and similar Mark 3 Desmo, now with a single filler gas tank and separate instruments. This is the Mark 3 D. *Ducati*

The Scrambler received a 450 motor for 1969 and a stronger frame. With few changes, the model lasted until 1975. One of the most popular wide-case singles, this is an early Euro-specification 450 Scrambler. A noticeable new feature was the large circular air filter on the right.

450 SCRAMBLER, MARK 3, MARK 3D
1969–1970 Differing from the 250 and 350 1968

Bore (mm)	86
Stroke (mm)	75
Displacement (cc)	435.7
Compression ratio	9.3:1
Carburetion	Dell'Orto VHB29AD
Power	27 horsepower at 6,500 rpm (SCR)
Dry weight (kg)	133 (SCR) 130 (Mark 3 & D)
Colors	Yellow, black stripe (SCR) Cherry Red (Mark 3 & D)
Engine numbers	From DM450 450001
Frame numbers	From DM450S 460001
Production 1969	2,507 (450 Scrambler) 359 (450 Mark 3) 418 (450 Mark 3 D)

250 and 350 Scrambler, Mark 3 and Mark 3 Desmo, 250 Monza, 160 Monza Junior

For 1969, the 350 Scrambler was marketed as the 350SSS (Street-Scrambler-Sport) in America. The engine was unchanged, still breathing through a racing-type Dell'Orto SSI carburetor, but this street emphasis made sales difficult for the Mark 3. On the Mark 3 and Desmo this year, a single filler fuel tank replaced the distinctive twin filler type, and the frame was painted black. As in 1968, some Mark 3s were also fitted with Scrambler fuel tanks, and the existing 250 Monza and 160 Monza Junior continued unchanged.

250 & 350 SCRAMBLER, M3, M3D, 250 160 MONZA 1969 Differing from 1968

Colors	Yellow, orange, red (SCR) Red / Cherry Red (Mark 3 & D)
Production 1969	2,190 (350 Scrambler) 2,536 (250 Scrambler) 213 (350 Mark 3D) 256 (350 Mark 3) 342 (250 Mark 3) 591 (250 Monza) 423 (160 Monza Junior)

For 1969, the 350 Scrambler was sold as the 350 SSS in America. Carburetion was still by the racing SSI.
Angus Dykman

PUSHROD SINGLES AND TWO-STROKES 1969
125 Cadet/4, 125 Cadet/4 Lusso, 125 Cadet/4 Scrambler, 100 Cadet, 100 Mountaineer

The existing pushrod singles continued unchanged, and the only two-strokes produced during 1969 were the 100 Cadet and 100 Mountaineer. While the 100/25 Brio and 50SL1 were still available in Italy, the local market was flooded with scooters and small capacity two-strokes, and the Ducatis weren't seen as anything special.

PUSHROD SINGLE AND TWO-STROKE PRODUCTION 1969	
440	(125 Cadet/4)
150	(125 Cadet/4 Lusso)
141	(125 Cadet/4 Scrambler)
310	(100 Cadet)
172	(100 Mountaineer)

1970

Although Ducati's sales were still relatively stagnant, during 1970 Milvio and Spairani provided Taglioni the resources to develop a pair of 90-degree V-twins: a racing 500 and production 750. Prototypes appeared during 1970, but in the meantime, Ducati's range continued much as before. A 50cc and 100cc two-stroke Scrambler were the only new models this year, and with demand for the 250, 350, and 450 Scrambler continuing to be strong, motorcycle production was a healthy 10,236.

OVERHEAD CAMSHAFT SINGLES 1970
250, 350, and 450 Scrambler 250, 350, 450 Mark 3 and Mark 3D, 160 Monza Junior

While the 450s continued unchanged, the 250 and 350 (Mark 3, Desmo, and Scrambler) received the new square-slide Dell'Orto carburetor sometime during 1970. The 250 Scrambler's carburetor was downsized to 26mm, with all other models receiving the 29mm unit. The Scrambler continued to be very popular, but the Mark 3 and Desmo languished. As in previous years, a few Mark 3s were fitted with Scrambler fuel tanks, but Berliner basically stopped importing the Mark 3 into America after 1970, deciding it was more economically prudent to concentrate on the new 450R/T and Spanish Mototrans models. This year also saw the end of the 160 Monza Junior, one of Ducati's biggest selling models of the 1960s.

PUSHROD SINGLES AND TWO-STROKES 1970
125 Cadet/4, 125 Cadet/4 Lusso, 125 Cadet/4 Scrambler, 100 Cadet, 100 Scrambler, 50 Scrambler

Although Berliner was no longer importing the range of pushrod singles and two-strokes, the Cadet continued on a limited scale. And in response to the then current fashion for dual-purpose motorcycles in Italy, Ducati released a pair of two-stroke Scramblers, these powered by versions of the 50SL and 100 Mountaineer engines. The 50 Scrambler was styled similarly to the four-stroke 125 Cadet Scrambler, while the 100 had a different gas tank and racing number side panels. The frame was the same dual-cradle type of the Cadet and SL, but the front fork was a more modern Ceriani type with exposed fork tubes. Both Scramblers

250, 350, 450 SCRAMBLER, MARK 3, MARK 3D, 160 MONZA JUNIOR 1970
Differing from 1969

Carburetion	Dell'Orto VHB26BD (250SCR) Dell'Orto VHB29AD (Mark 3, D, 350SCR)
Colors	White, yellow, blue (Mark 3 & D)
Engine numbers	To DM250 108620 (250) DM350 11052 (350) DM450 453508 (450)
Production	2,511 (450 Scrambler) 2,495 (350 Scrambler) 2,537 (250 Scrambler) 206 (450 Mark 3 D) 199 (450 Mark 3) 110 (350 Mark 3D) 133 (350 Mark 3) 157 (250 Mark 3D) 178 (250 Mark 3) 200 (160 Monza Junior)

50, 100 SCRAMBLER 1970

Engine	Single-cylinder piston-port two-stroke, air-cooled
Bore (mm)	38.8 (52)
Stroke (mm)	42 (46)
Displacement (cc)	49.66 (97.69)
Compression ratio	10.5:1 (11.2:1)
Carburetion	Dell'Orto UBF24BS
Power	3.27 horsepower (6.27 hp)
Gears	Four-speed
Front suspension	Telescopic fork
Rear suspension	Twin shock swingarm
Brakes (mm)	Drum 118
Tires	2.50x18, 2.50x17 (2.75x18, 3.75x16, 100)
Wheelbase (mm)	1,150 (1,180)
Dry weight (kg)	70 (71)
Production	315 (50) 500 (100)
Other pushrod single and two-stroke production 1970	257 (125 Cadet/4) 108 (125 Cadet/4 Lusso) 90 (125 Cadet/4 Scrambler) 240 (100 Cadet)

The 1960's final small-capacity two-strokes were the uninspiring 50 and 100 Scrambler. This is the 100 Scrambler. *Ducati*

included a high-rise exhaust and a high front fender, and although the 50 included a metal grille over the headlight, these were not really serious off-road motorcycles. The 50 and 100 Scramblers weren't particularly popular, and by 1970 Ducati's new management decided the future lay in more profitable larger displacement motorcycles. The two-strokes eventually died, for a short while at least.

Looking back on the decade of the 1960s, Ducati was undoubtedly lucky to survive. In an era dominated by poor decisions and the production of a number of uninspiring models at the dictate of Berliner, Ducati was very fortunate it was not economically accountable. Unlike the Japanese who marched into America with a clear plan, Berliner didn't seem to have any real grasp of the US motorcycle market, but fortunately the new management at Borgo Panigale initiated a change in direction. And with a 750cc Superbike on the agenda, Ducati looked to put that difficult decade behind it.

CHAPTER 4

DUCATI'S FIRST SUPERBIKE
THE 750, SINGLES, AND TWINS 1971–1974

Spairani, Calcagnile, Milvio, and Taglioni were the four protagonists responsible for the 750 Ducati. Here they are with one of the preproduction examples in July 1971. *Ducati*

During the 1960s, Ducati gained an enviable reputation for producing high-quality sporting singles, but it was the transition to the manufacture of larger capacity machines that set the company on the path of glory that continues today. Developed concurrently with the 500 Grand Prix twin, the 750GT would be the most significant new model since the Marianna. Prior to the 750, the bevel-drive overhead camshaft single defined Ducati, but with the 750 this became the 90-degree V-twin. While the 750 was initially treated with skepticism, the decisive victory in the 1972 Imola 200-mile race was a pivotal event in Ducati's history, giving the fledgling 750 credibility in a market then saturated with new Superbikes from Japan and Europe. Imola led to the 750 Sport and 750 Super Sport, and soon a whole range of bevel-drive twins.

1971

Although Ducati was still struggling with strikes and uncertainty with component supply, 1971 saw the release of several important new models, these representing a significant change in focus. While the new wide-case single was ostensibly a continuation of an earlier theme, with the 450R/T and new Desmo singles, Ducati moved toward a more off-road and street sporting emphasis. At the same time it embarked on an official Grand Prix racing program for the first time since 1958. Although motorcycle production dipped slightly, to 9,916, the concentration was now on more saleable models, in particular the range of four-stroke Scramblers.

Colin Seeley created a special frame for the 500 Grand Prix Ducati. Here he is with the result in early 1971. *Ducati*

500GP 1971–1972

Engine	90-degree V-twin four-stroke, air-cooled
Bore (mm)	74
Stroke (mm)	58
Displacement (cc)	498
Compression ratio	10:1 (8:1)
Valve actuation	Bevel gear–driven single overhead camshaft
Power	61.5 horsepower (1971) 72 horsepower (1972)
Gears	Six-speed
Front suspension	35mm Marzocchi fork
Rear suspension	Twin Ceriani shock absorber swingarm
Tires	3.00x18, 3.25x18
Wheelbase (mm)	1,430
Dry weight (kg)	135

Giuliano on his way to fourth on the 500GP at Sanremo Ospedaletti in October 1971. *Ducati*

500 Grand Prix Twin and 350 Triple

Alongside the 750, Taglioni spent much of 1970 working on a 500cc Grand Prix V-twin. Hoping to match Giacomo Agostini's 500cc MV Agusta triple, the basic engine layout was similar to the 750 but was considerably more compact. To speed up development, Spairani commissioned Colin Seeley to build a racing frame. Marzocchi provided special leading axle forks with a single Lockheed front disc brake, and at the rear were Ceriani shock absorbers and a 200mm Fontana double leading shoe drum brake. Ducati decided introduce the 500 by contesting some rounds of the 1971 Italian Senior Championship and Italian Internationals, Spaggiari and Ermanno Giuliano achieving some respectable results but never managing to defeat Agostini.

Another of Milvio and Spairani's initiatives during 1971 was the development of a three-cylinder 350cc Grand Prix engine. Unlike the V-twin, this wasn't designed in-house, the work entrusted to Ricardo in England. Although a prototype engine was built, the project was scrapped during 1972.

750GT

The 750GT was released in July 1971, and as Taglioni needed to be economically pragmatic, the new 750 V-twin (or L-twin as he called it) inherited many design characteristics from the wide-case overhead camshaft single. The aluminum crankcases were still vertically split but with the front cylinder inclined upwards 15 degrees. The crankcases on the 1971 engines were sand-cast, and as on the single, the crankshaft was a pressed-up assembly with the connecting rods supported by roller bearings. Also inherited from the single was the wide included valve angle of 80 degrees, arguably obsolete by 1971, and to keep the rocker box as narrow as possible, single coil valve springs replaced the hairpin valve springs of the single.

ABOVE: The first series 750GT was produced as a preproduction series, with many specific details differing from later versions.

LEFT: The 750GT heralded a new age for Ducati, and the 90-degree L-twin established a new engine blueprint.

Lubrication for the 750 engine was by wet sump, and the only oil filtration was a gauze filter in the sump. The electrical system was upgraded to 12 volts, and ignition was by battery, coil, and points. To minimize engine length, the constant-mesh direct-type five-speed gearbox had the layshaft above the mainshaft. Carburetion was by a pair of British Amal concentric carburetors, and air filtration consisted of two separate black plastic airboxes, each with a dry paper element, fed via a restrictive pleated plastic tube. The barely street-legal exhaust system included twin Conti mufflers.

The 750cc V-twin was housed in a Seeley-inspired tubular steel frame that used the engine as a stressed member in the usual Ducati fashion. An improvement over the wide-case single was tapered roller steering head bearings. The Marzocchi front fork had a leading axle, with a single Lockheed front disc and twin-leading shoe rear drum brake. The wheels were aluminum Borrani, and the 1971 750GTs had a metalflake fiberglass 17-liter fuel tank, stainless-steel fenders, and a number of subtle differences that set them apart from later examples. These included footpegs positioned further forward, specific Smiths instruments, and a thicker saddle. Designed by Taglioni, Renzo Neri, and Franco Farnè, the metalflake styling evolved from the contemporary silver Desmo single.

250, 350, and 450 Desmo

For 1971, Ducati transformed the Mark 3 Desmo into one of the definitive sporting Ducati singles, the Desmo 250, 350, and 450. Restyled by Taglioni, Neri, and Farnè, the

750GT 1971

Engine	90-degree V-twin four-stroke, air-cooled
Bore (mm)	80
Stroke (mm)	74.4
Displacement (cc)	748
Compression ratio	8.5:1
Valve actuation	Bevel gear–driven single overhead camshaft
Carburetion	30mm Amal R 930/76 and /77
Power	57 horsepower at 7,700 rpm
Front suspension	38mm Marzocchi fork
Rear suspension	Marzocchi 305mm twin shock swingarm
Front brake (mm)	275 single-disc
Rear brake (mm)	200 drum
Wheels	Borrani WM2x19 front, WM3x18 rear
Tires	3.25x19, 3.50x18
Wheelbase (mm)	1,530
Dry weight (kg)	185
Top speed (km/h)	200
Colors	Silver or black frame Metalflake blue, orange, green, red
Production	660 30 (USA)
Engine numbers	From DM750 750001
Frame numbers	From DM750S 750001

Nicknamed "The Silver Shotgun," the 1971 350 Desmo established a new sporting style for Ducati.

The front fork and brake were improved over earlier Desmos.

Desmo's focus was now that of a true café racer, with clip-on handlebars, rear-set footpegs, and the absolute minimum of street equipment. With its garish silver metalflake fiberglass bodywork, the 1971 and 1972 Desmo made more of a statement than its predecessor, earning the nickname "The Silver Shotgun."

The first of the new Desmo singles was the 350, the 350cc Desmo engine continuing as before except for a larger right side main bearing. Although some early examples had a Dell'Orto SSI carburetor, most production versions had the usual Dell'Orto VHB. While the frame was similar to the Mark 3

250, 350, 450 DESMO
Differing from the 250, 350 and 450 Mark 3 Desmo

Power	26 horsepower at 9,000 rpm (250)
	29 horsepower at 8,500 rpm (350)
	31 horsepower at 6,500 rpm (350)
Front suspension	35mm Marzocchi fork
Wheels	Borrani WM2x18 front and rear
Tires	2.75x18, 3.25x18
Dry weight (kg)	128
Top speed (claimed km/h)	150 (250)
	165 (350)
Colors	Metalflake silver
Engine numbers	From DM250 110000 Approx.
	From DM350 12200 Approx.
	From DM450 455500 Approx.
Frame numbers	From DM250GT 102500 Approx.
	From DM350S 355000 Approx.
	From DM450M3 700100 Approx.
Production	380 (450)
	170 (350)
	300 (250)

The minimalist instrumentation included only a large central tachometer on this example. Most had a speedometer incorporated in the headlight shell.

Desmo, the front suspension, front brake, and wheels were new. The Marzocchi front fork was now a beefier 35mm unit, the front brake a double-sided single leading shoe Grimeca, and the wheels aluminum Borrani. Setting the 1971 Desmo apart was new silver metalflake fiberglass bodywork, while the minimalist sporting style extended to clip-on handlebars and rear-set footpegs, with a large white Veglia tachometer dominating the cockpit.

As a sporting motorcycle, the Desmo was functional and without pretension, and soon after the 350 was released, identical-looking 250 and 450 versions were available. As the silver Desmo wasn't produced in significant numbers, and not even sold in the United States, they are now rare and considerably desirable. The performance may have been underwhelming, but handling was impeccable and they brilliantly conveyed Ducati's evolving sporting ethos.

350, 450 R/T

While the 450 Scrambler was a successful attempt at creating a dual-purpose motorcycle with street orientation, Berliner wanted a more effective dirt motorcycle to take on the BSA 441 Victor in America. The resulting 450R/T was quite different to the Scrambler and sold in the United States as an off-road machine, with a kit providing street equipment.

The 450R/T engine was the same desmodromic unit as fitted to the 450 Desmo, but the ignition was by a flywheel magneto and there was no battery or ignition key. Optional street equipment included an Aprilia headlight and CEV taillight on a rubber bracket. A completely new frame was designed for the R/T, sharing several features with the 750GT unit developed at the same time. Although retaining a single downtube like the other singles, horizontal tubes from the subframe reinforced the backbone and steering head, with a bolted-on rear section.

The 450R/T's Desmo engine also included a decompression lever behind the bevel gear shaft.

350, 450R/T
Differing from the 350 and 450 Desmo

Brakes (mm)	Drum 160 front and rear
Tires	3.00x21, 4.00x18 Pirelli Cross
Wheelbase (mm)	1,450
Dry weight (kg)	128
Colors	Black or silver frame, yellow
Engine numbers (450R/T)	From DM450 453000 Approx.
Production	6 (350R/T)
	194 (450R/T)

All the 450R/T's bodywork was fiberglass and the wheels lightweight Borrani.

ABOVE RIGHT: The Desmo 450R/T was a more serious off-road motorcycle than the Scrambler. This Italian specification example has a dual muffler and street equipment.

The steering head bearings were also tapered rollers like the 750, but the swingarm with its cam washer rear wheel adjuster was unique to the R/T. The suspension included a 35mm Marzocchi fork with polished aluminum legs and exposed tubes providing 7 inches of travel.

Separating the R/T from the Scrambler was the 21-inch aluminum Borrani front wheel and smaller front brake, 10-liter fiberglass gas tank, and rubber-mounted fiberglass fenders. Widely promoted and heavily advertised by Berliner, the 450R/T was an impressive effort but unfortunately not competitive with the new wave of 400cc off-road two-strokes. A few similar 350R/Ts were also built this year.

250, 350, 450 Scrambler and Mark 3

Except for small updates, the 1971 Scrambler continued as it was for 1970. As the coffin tank Mark 3 proved unpopular and the Scrambler tank version looked very similar to the Scrambler, the Mark 3 was restyled with a new, larger fuel tank for 1971. The engines were shared with the Scrambler, and apart from the new 4-gallon Diana-shaped tank and a thicker seat, the rest of the Mark 3 was unchanged and still included the skinny old-style 31.5mm Ducati fork and steel wheels. As the Mark 3's style now deviated from the more sporting Desmo, clip-on handlebars were no longer offered, the handlebar was an upright touring style more in keeping with the forward mount footpegs and rocking gear shift lever.

The 1971 Mark 3 was produced mainly for the Italian market and wasn't imported in any quantity into the United States. Also joining the 1971 lineup in Italy was the lower compression touring 450T and 450TS. Specific touring equipment included a deep front fender, a pair of painted steel panniers, and a crash bar. The most significant update was the introduction of a 12-volt electrical system.

250, 350 AND 450 SCRAMBLER AND MARK 3, 450T, TS
Differing from 1970

Rear tire	3.25x18 (350 and 450 Mark 3)
Engine numbers	From 108620 (250) From 11051 (350) From 453509 (450)
Production	2,502 (450 Scrambler) 2,302 (350 Scrambler) 1,956 (250 Scrambler) 284 (450 Mark 3) 222 (350 Mark 3) 433 (250 Mark 3) 34 (450 TS)

CLOCKWISE FROM TOP LEFT:

The 1971 350 Scrambler had a new taillight and instruments. *Ducati*

Although the Scrambler was little changed, Ducati revamped its image for 1971.

One of the least successful models of the early 1970s was the touring 450TS.

The Mark 3 was moderately restyled. This is the 450, but the 250 and 350 were similar. *Ducati*

125 Scrambler, 100 and 50 Scrambler

While production of the narrow-case overhead camshaft engine finished in Bologna, Ducati subsidiary Mototrans in Spain supplied 125cc motors for the 125 Scrambler in 1971 and 1972. Specifications included a Spanish Amal Monobloc carburetor and a five-speed gearbox. Unlike other overhead camshaft singles, the frame was a duplex full cradle type. Most of the other components were Italian, including Marzocchi suspension and CEV instruments and taillight. While a few were sold in the United States, the expensive 125 Scrambler wasn't especially popular. Also continuing unchanged this year was the 100 and 50 two-stroke Scrambler.

1972

Although the new 750 was in production and receiving good reviews, the delayed construction of a new building severely disrupted production into 1972. With considerable industrial unrest and intermittent rolling stoppages, it wasn't until March that the metal workers' agreement was signed and the factory could operate for the rest of the year without the threat of strikes. The turning point came in April, with a spectacular victory in the inaugural Imola 200-mile race. After Imola, Ducati's 750 was seriously treated as a leading Superbike and Ducati was no longer a minor manufacturer of esoteric singles with strange valve operation.

Unfortunately, Ducati wasn't in a position to meet the demand for the 750 following Imola and ultimately 8,546 motorcycles of all capacities were produced during 1972. Development

125 SCRAMBLER 1971

Engine	Single-cylinder four-stroke, air-cooled
Bore (mm)	55.2
Stroke (mm)	52
Displacement (cc)	124.443
Compression ratio	8.5:1
Valve actuation	Bevel gear-driven single overhead camshaft
Carburetion	Amal 375/20
Power	10 horsepower at 8,000 rpm
Gears	Five-speed
Front suspension	Marzocchi fork
Rear suspension	Twin shock swingarm
Brakes (mm)	Drum 158 and 136
Tires	2.50x19, 3.50x18
Wheelbase (mm)	1,340
Dry weight (kg)	105
Colors	Red, yellow, orange, black frame
Engine numbers	From DM 20001–21000 Approx.
Production	100 (125 Scrambler) 60 (USA 125 Scrambler) 160 (50 Scrambler) 120 (100 Scrambler)

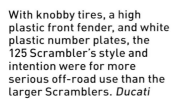

With knobby tires, a high plastic front fender, and white plastic number plates, the 125 Scrambler's style and intention were for more serious off-road use than the larger Scramblers. *Ducati*

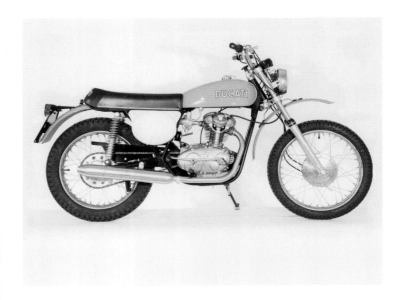

was limited to the introduction of the 750 Sport, with all the existing 750GT and singles continuing ostensibly unchanged.

The Imola 200

With the announcement of the Imola 200 "Daytona of Europe" to be held on April 23, 1972, Spairani instructed Taglioni to mount a full-scale attempt at winning the race. Riders were Paul Smart, Bruno Spaggiari, Ermanno Giuliano, and Alan Dunscombe, and although the racing desmodromic 750s looked surprisingly similar to the 750GT, they were highly developed factory racers sharing little with the production 750. The engine included lighter and stronger con rods, higher compression pistons, high-lift desmodromic camshafts, and a pair of new concentric Dell'Orto 40mm carburetors. Ignition was by twin spark plugs per cylinder. The frame began life as a production unit, retaining center stand mounts, while the fiberglass fuel tank included a large clear stripe as an instant fuel gauge.

On race day, 70,000 spectators crammed in to see who would win the record total prize money of 35,000,000 lire. Alongside the factory Ducatis were the MV Agustas of Giacomo Agostini and Alberto Pagani, and Moto Guzzis of Guido Mandracci and Jack Findlay. Other factory teams included Norton, BSA, and Triumph, with semi-factory Kawasakis, Hondas, and BMWs completing the field. When the race began, the two silver Ducatis initially followed Agostini, and after Agostini's retirement, Smart and Spaggiari circulated together, Smart eventually crossing the line four seconds ahead of Spaggiari. Smart's race average was 157.353 kilometers an hour, and he shared the fastest lap of 161.116 kilometers an hour with Spaggiari and Agostini. As Ducati had proven to the world, its desmodromic 750 could take on all comers and win; this was a pivotal victory in Ducati's history.

F750 IMOLA 1972 Differing from the 750GT

Compression ratio	10:1
Valve actuation	Bevel gear–driven desmodromic single overhead camshaft
Carburetion	Dell'Orto PHM40
Power	86 horsepower at 9,200 rpm
Front brake (mm)	275 twin-disc
Rear brake (mm)	230 disc
Wheels	Borrani WM3x18 front and rear
Tires	3.25x19, 3.50x18
Dry weight (kg)	178
Top speed	272 km/h (169 mph)
Color	Metalflake silver

CLOCKWISE FROM TOP LEFT:

The impressive lineup of factory 750s at Imola in 1972. *Ducati*

Paul Smart led Bruno Spaggiari home in a spectacular one-two result. *Ducati*

Fabio Taglioni with Paul Smart after the Imola victory. *Ducati*

DUCATI'S FIRST SUPERBIKE | 109

ABOVE: The 90-degree 750GT engine layout established a defining configuration for Ducati engines.

ABOVE RIGHT: The 1972 750GT was similar to the 1971 version, but the fiberglass bodywork was generally no longer metalflake.

750GT AND 750 SPORT

As the 750GT entered regular production, numbers increased (to 1,062 this year), and it was updated slightly. A specific US version was now also available, this including higher handlebars, turn signals, and a larger taillight. Joining the 750GT during 1972 was a more sporting 750 Sport. Based on the GT, the first 750 Sport still featured the GT frame with wide rear subframe but included a higher performance engine with black outer covers. Inside the engine were higher compression pistons, and carburetion was by a pair of the new larger Dell'Orto concentric carburetors with accelerator pumps. Because these breathed through open bell mouths, the 750 Sport's performance was increased significantly over the GT. Completing the sporting specification were clip-on handlebars, rear-set footpegs, and minimalist instrumentation. With its distinctive yellow and black fiberglass bodywork, the 1972 Sport is known as the Z-stripe.

The 1972 750GT is still a very handsome machine.

750 SPORT 1972
Differing from the 750GT

Compression ratio	9.3:1
Carburetion	Dell'Orto PHF32A
Power	62 horsepower at 8,200 rpm
Dry weight (kg)	182
Top speed (km/h)	210
Colors	Yellow and black
Production	645 (750 Sport) 90 (USA 750 Sport) 642 (750GT) 150 (USA 750GT) 270 (Electric Start 750GT)
Engine numbers	From DM750 750700 Approx.
Frame numbers	From DM750S 750700 Approx.

250, 350, and 450 Scrambler; Desmo; Mark 3; 450 R/T; 125 Scrambler

With production delays and the 750 dominating the market, the existing range of overhead camshaft singles continued unchanged for 1972. Two-stroke Scrambler production went into hiatus but would recommence during 1973. While the overhead camshaft Scramblers may have been overshadowed by the 750 and sporting singles, with 4,940 built, they continued as the most popular model in the range.

PRODUCTION OHC SINGLES 1972

163 (450 R/T USA)
430 (450 Mark 3)
400 (350 Mark 3)
380 (250 Mark 3)
1,561 (450 Scrambler)
1,368 (350 Scrambler)
2,011 (250 Scrambler)
158 (125 Scrambler)

ABOVE: 450 Scramblers awaiting delivery at the Borgo Panigale factory in October 1972. *Allan Tannenbaum*

BELOW: The first 750 Sport looked striking and had a wide rear subframe and Z-stripe decals.

DUCATI'S FIRST SUPERBIKE | 111

1973

Although Ducati was still basking in the Imola success, sales remained stagnant and EFIM considered Ducati's management financially irresponsible. During 1973, a new director, Cristiano De Eccher, replaced Milvio and Spairani, to make the Borgo Panigale plant more profitable. De Eccher immediately curtailed the racing program, sanctioning only two events (Imola and the Barcelona 24-hour race), and initiated the demise of the 750 and overhead camshaft singles, both deemed too expensive to manufacture. As Ducati increased the prices of the overhead camshaft singles, Berliner chose not to import them, instead sourcing the 250 and 350 from Mototrans in Spain.

Despite this rather gloomy outlook, production at Borgo Panigale remained healthy, with 7,632 motorcycles manufactured. And Ducati was still very successful on the track, Bruno Spaggiari finishing second in the Imola 200 on a special short-stroke 750 and Benjamin Grau and Salvador Canellas winning the Barcelona 24-hour race at Montjuïc on a prototype 860.

Ducati fielded three special short-stroke racers in the 1973 Imola 200. Spaggiari (No. 84) finished second. The others were ridden by Bruno Kneubuhler (No. 86) and Mick Grant (No. 85). Taglioni is behind. *Ducati*

The 1973 750 Super Sport was based on the 750 Sport and had black engine covers. This was the *Cycle* magazine racing bike of 1975–1976, since restored to original street specification.

750 Super Sport, 750 Sport, 750GT

Immediately after the Imola victory, Spairani promised production Imola Replicas, but the first of these didn't appear until mid-1973. The first 750 Super Sports were initially modified 750 Sports, and while only a few were built, one went to *Cycle* magazine in California where Cook Neilson and Phil Schilling subsequently developed it into a competitive Superbike.

Setting the Super Sport engine apart from the Sport were the desmo heads, 40mm carburetors, and special con rods, while the chassis included triple disc brakes, a half fairing, and a large fiberglass tank with the trademark clear-stripe instant fuel level gauge. While the 750SS styling was a joint effort between Taglioni, Neri, Farnè, and Ducati's racing subsidiary NCR, the 750 Sport was restyled by Leopoldo Tartarini for 1973, now with a narrow rear subframe and new decals for the fiberglass bodywork. The 750GT gained a steel gas tank and side covers, and an Italian Scarab front brake replaced the British Lockheed, but in other respects the specifications were unchanged.

750GT AND SPORT 1973 Differing from 1972

Colors	Orange/black, red/black (750GT)
Engine numbers	From DM750 752200 Approx.
Frame numbers	From DM750S 752200 Approx.
Production	1,817 (750GT)
	746 (750 Sport)

DUCATI'S FIRST SUPERBIKE

Long, lean, and narrow, the 1973 was pure sportster.

One of Ducati's most impressive styling efforts of any era: the 1973 750 Sport.

For 1973, the 750GT gained a steel gas tank and side covers. Euro-spec versions like this had black fork legs and rear shock covers.

250, 350, and 450 Scrambler, Desmo, Mark 3, 50 Scrambler

Updates to the range of overhead camshaft singles saw the adoption of Ducati Elettrotecnica or Motoplat electronic ignition and a new headlight and instruments. The Scrambler and Mark 3 gained a 35mm Marzocchi front fork, Borrani wheels, and the double-sided front brake, while the Desmo was restyled (also by Tartarini) to replicate the 750 Sport. This included a new one-piece fiberglass seat/tail piece. The two-stroke 50 Scrambler also made a brief return.

ABOVE RIGHT: 1973 750 GT engines at the factory toward the end of 1972. Most GTs this year had an electronic tachometer and no mechanical drive off the front cylinder. Note the factory lead seal underneath the crankcase. *Allan Tannenbaum*

RIGHT: The 1973 250 Scrambler. This now had an engine supplied by Mototrans in Spain. *Ducati*

BELOW: Tartarini supplied new styling for the 1973 Desmo single. Retained this year was the drum front brake.

DUCATI'S FIRST SUPERBIKE | 115

The 450 Mark 3 was also updated for 1973, now with the Desmo's wheels, brakes, and suspension, along with attractive blue colors.

1974

This was a watershed year for Ducati. Production of the round-case 750 ended, with the venerable overhead camshaft single following soon afterward, and by the end of the year, Ducati headed in a completely new direction. In the midst of this uncertainty, Ducati released arguably its most significant production model, the 750 Super Sport. But despite 750 and single-cylinder production winding down, a healthy 7,703 motorcycles were produced this year. It was ironic that just as Ducati's management saw no future in the 750 and single, they became more popular. Now they are considered some of Ducati's most desirable models.

250, 350, 450 SCRAMBLER; MARK 3; DESMO; 50 SCRAMBLER
Differing from 1972

Colors	Yellow (Desmo)
	Blue (Mark 3)
Engine numbers	250, 350, 450M3, or
	250, 350, 450D
Frame numbers (350 Desmo)	From DM350D 400002
Production	330 (450 Desmo)
	389 (350 Desmo)
	300 (250 Desmo)
	350 (450 Mark 3)
	476 (350 Mark 3)
	353 (250 Mark 3)
	1,336 (450 Scrambler)
	1,156 (350 Scrambler)
	200 (250 Scrambler)
	179 (50 Scrambler)

750 Super Sport

After a two-year wait, the Imola Replica Desmodromic 750 Super Sport finally became available in limited numbers during 1974. At the time, it was one of the most exotic production motorcycles available, a true race replica boasting triple disc brakes, a racing fairing, and an engine prepared under the watchful guidance of none other than Fabio Taglioni.

Unlike the 750 Sport, the Super Sport engine was substantially upgraded over the 750GT. The crankshaft included milled and machined con rods, with dual strengthening ribs around the big-end eye. Carburetion was by a pair

of Dell'Orto 40mm carburetors, and inside the desmodromic cylinder heads were highly polished valve rockers. While the frame was similar to the 750 Sport, the center axle Marzocchi fork and 18-inch front wheel were new. Braking was by twin Scarab front disc brakes and a single Lockheed disc at the rear. The 1972 Imola racers inspired the sculptured 20-liter fiberglass gas tank with clear fuel gauge stripe and solo racing saddle with zipper compartment.

The desmodromic round-case 750 had only one series produced in 1974. While they were flawed, with dubious brakes and fiberglass, they were rare, beautiful, exotic, and functionally superior. The round-case 750 Super Sport has justifiably earned a place as the most coveted production Ducati.

ABOVE LEFT: Although released two years after the Imola 200 victory, the 750 Super Sport Imola Replica was worth the wait.

ABOVE RIGHT: Inside the 750 Super Sport's engine were desmodromic valves, polished rockers, and special milled con rods.

BELOW: Hand-built in limited numbers, the 750 Super Sport has become one of the most coveted production Ducatis.

750 SUPER SPORT 1974
Differing from the 750 Sport

Compression ratio	9.5:1
Valve actuation	Bevel gear–driven desmodromic single overhead camshaft
Carburetion	Dell'Orto PHM40A
Power	70 horsepower at 9,000 rpm
Front brake (mm)	275 dual-disc
Rear brake (mm)	230 disc
Wheels	Borrani WM3x18 front and rear
Tires	3.50x18, 3.50x18 C7 Metzeler Racing
Wheelbase (mm)	1,500
Dry weight (kg)	180
Top speed (km/h)	230
Colors	Blue/green frame, silver
Engine numbers	DM750.1 075001–075411
Frame numbers	DM750SS 075001–075411
Production	401

DUCATI'S FIRST SUPERBIKE | 117

750GT AND SPORT 1974
Differing from 1973

Front suspension	Ceriani or Marzocchi center-axle
Front brake	Brembo
Engine numbers	From DM750 754000 Approx.
Frame numbers	From DM750S 754000 Approx.
Production	1,675 (750GT) 856 (750 Sport)

RIGHT: The 750 Sport gained a steel gas tank and polished engine covers for 1974.

For 1974, the 750 GT gained Dell'Orto carburetors and a center-axle fork. Most had a Brembo brake like this. New this year were the painted fenders.

750 Sport and 750GT

For their final year of production, the 750 Sport and 750GT evolved into slightly less pure versions but arguably more functional. The 750 Sport gained a steel fuel tank, polished engine cases, and a center-axle fork, mostly with an improved Brembo brake, but was compromised by new CEV switches. The 750GT was similarly afflicted, marred by the end of its production with ugly Lafranconi mufflers and steel Radaelli wheel rims. Fortunately, the soul of both these magnificent models remained intact, but production would end during 1974.

239, 250, 350, 450 SCRAMBLER, MARK 3, DESMO, R/T 1974
Differing from 1973

Front Brake	280mm Disc (Desmo)
Production	454 (450 Desmo)
	595 (350 Desmo)
	427 (239/250 Desmo)
	263 (450 Mark 3)
	324 (350 Mark 3)
	333 (239/250 Mark 3)
	73 (450 Scrambler)
	331 (350 Scrambler)
	200 (250 Scrambler)
	100 (350, 450 R/T)

239, 250, 350, and 450 Scrambler Desmo, Mark 3, R/T

With their days numbered, the overhead camshaft Scrambler and Mark 3 continued much as before, while the Desmo single gained a Ceriani fork and Brembo front disc brake. This year also saw a limited number of 239 Mark 3 and Desmos produced, primarily for an anomaly in French law that allowed for lower taxes on motorcycles under 240cc.

The final Desmos were an enigma. While the electrical system was still 6 volts, and the kick-start inconveniently still on the left, Ducati attempted to bring the single in the 1970s with electronic ignition and a front disc brake. On one hand, the final Desmos were a relic from the past, but on the other they were thoroughly modern. Ultimately, as the bevel-drive single-cylinder engine was almost as expensive to manufacture as a twin, economic reality eventually killed the overhead camshaft single.

New for the Desmo single for 1974 was a Ceriani front fork and Brembo disc brake.

The 1974 250 Desmo marked the end of the line of Ducati sporting singles.

DUCATI'S FIRST SUPERBIKE | 119

CHAPTER 5

ECONOMIC RATIONALIZATION
THE 860, 900, PARALLEL TWIN, AND PANTAH, 1975–1984

Cook Neilson provided Ducati its most significant victory yet, comfortably winning the 1977 Daytona Superbike race on a 750SS.

Grau and Canellas provided Ducati another victory in the Barcelona 24-hour race during 1975. *Ducati*

The 1960s and 1970s were characterized by a number of dubious production and marketing decisions, and 1975 in particular was a low point. As had happened during the grim 1960s, Ducati's management again saw their future in America, with the new 860GT replacing the expensive 750 and a range of parallel twins replacing the long-running overhead camshaft singles. Considering none of Ducati's earlier parallel twins were successful, this was a surprising move, and it ultimately proved a mistake. Completing a dubious new lineup was the reintroduction of a two-stroke off-road model, the 125 Regolarità, for Italy and France. While it seemed Ducati had lost its way, the company managed some salvation with the release of the new 900 Super Sport, and this would sustain sales for the rest of the decade.

1975

Although Ducati was no longer officially involved in racing, this year again saw Ducati triumph in the Barcelona 24-hour race, where Grau and Canellas repeated their 1973 victory, winning at an average speed of 71.4 miles per hour on the tight Montjuïc circuit. And in anticipation of a strong sales year driven by an all-new model lineup, production slipped only slightly, to 5,977 motorcycles built in 1975.

860GT and 860GTE

While the 750 round-case was becoming prohibitively expensive to manufacture, several other factors also hastened its demise. New US DOT regulations coming into effect in September 1974 required a mandatory left-side gearshift and the 750's barking Conti exhaust simply wasn't street legal in most countries. To maintain performance with more stringent mufflers, the simplest solution was to increase displacement. As with the earlier singles, the design of the 750 allowed an easy displacement increase, and it was a relatively simple process to install 450 single pistons in the 750 to give 860cc. While an 860 was long anticipated, the eventual form took many by surprise. Rather than continue the sporting emphasis of the 750, the 860 placed styling first, along with a revamp of the engine and chassis.

To make the 90-degree twin more attractive for the US market, the 750GT was redesigned and restyled. The engine underwent considerable modification, first to simplify manufacture, and secondly to improve reliability and minimize maintenance. Taglioni was instructed to redesign the 750 engine so that it was much less labor intensive to assemble and cheaper to manufacture. At the same time, Giorgetto Giugiaro at Studio Italdesign in Turin was given the styling brief, and the outer engine cases were redesigned to match the angular 860 styling.

There was a lot more to the engine redesign though than merely the shape of the outer engine covers. Apart from the gearbox, most of the internal parts were not interchangeable with the 750. The bevel-gear layout to drive the two vertical shafts now included spur gears mounted onto an outrigger plate, and it took considerably less time to assemble. Other improvements included an oil filter between the cylinders and Ducati Elettrotecnica electronic ignition. The exhaust system included Lafranconi mufflers, styled to blend with the 860 tank and seat, and the gearshift was crudely moved to the left side of the engine via a rod running behind the engine. Unlike the 750GT, an electric starter was integral to the 860 concept, although early examples were kick-start only.

Most of the 860GT/GTE chassis was new. The swingarm included eccentric pivot chain adjusters, and the front wheel was 18 inches in diameter. A big effort was made with the 860 to improve many components that had come under criticism on the 750, and the CEV switches and electrical system were completely new. Production of the 860 commenced in September 1974, for the 1975 model year, but the styling didn't meet universal acclaim and reliability was dubious, particularly the electronic ignition. Now the 860GT is seen as being ahead of its time, the angular styling more representative of the late 1970s and early 1980s.

An electric start was integral to the 860 design.

With its new angular styling by Giugiaro, the 860 represented a significant departure from the round-case 750.

750 and 900 Super Sport

When the 860GT was conceived in late 1973, there was no intention of creating a Super Sport version. Limited-edition race replicas were uneconomic to produce and difficult to justify. But the 860GT's initial failure saw another limited production run of Super Sports during 1975. Arguably the finest of all the Super Sports, they flew in the face of legislation by retaining a right-side gearshift, Conti mufflers, and unfiltered 40mm carburetors. Not legally saleable in the United States, this small series was so successful that it led to the regular production of both the 750 and 900 Super Sport during 1976, adapted especially for America.

The 1975 900 Super Sport was one of Ducati's finest efforts, the soft flowing lines of the fairing, tank, and seat contrasting with the hard-edged angular engine cases.

860GT, 860GTE, 1975
Differing from the 750GT

Bore (mm)	86
Displacement (cc)	863.9
Compression ratio	9:1
Carburetion	Dell'Orto PHF32A
Power	57 horsepower at 7,700 rpm
Front suspension	38mm Ceriani fork
Rear suspension	Marzocchi 320mm twin shock swingarm
Front brake (mm)	280mm single- or twin-disc
Wheels	Radaelli WM3x18 front and rear
Tires	3.50x18, 3.50x18 (120/90x18)
Wheelbase (mm)	1,550
Dry weight (kg)	185
Top speed (km/h)	195
Colors	Black frame, orange, red, blue, black, yellow, green
Engine numbers	From DM860 850001
Frame numbers	From DM860S 850001
Production (860GT & 860GTE)	1,671 (1974) 1,316 (1975)

750 AND 900 SUPER SPORT 1975
Differing from the 1974 750SS

Bore (mm)	86 (900)
Displacement (cc)	863.9 (900)
Compression ratio	9.65:1 (750), 9.5:1 (900)
Rear suspension	Marzocchi 310mm twin shock swingarm
Front brake (mm)	280 dual-disc
Rear brake (mm)	229 disc
Colors	Silver frame, silver and blue
Engine numbers	From DM750.1 075412 (750SS) DM860.1 086001 (900SS)
Frame numbers	From DM750SS 075412
Production	249 (750SS) 246 (900SS)

Raw and purposeful, the 1975 Super Sport continued the style of the famed 1974 "Green-Frame" 750 Super Sport.

BELOW RIGHT: Styled to mimic the 860GT, the 500GTL replaced the charismatic overhead camshaft singles and wasn't a success.

Produced in 750 and 900cc versions, the square-case engine was derived from the 860GT. The chassis was essentially that of the earlier 750SS, and the result was an intoxicating blend of the high-torque square-case engine in the excellent 750 chassis. Like the round-case 750SS, the 1975 examples were also hand-built limited editions, and with no provision for frame-mounted turn signals, the 1975 Super Sport represented the end of an era. Although afflicted with the suspect Ducati Elettrotecnica ignition system and weak bottom end, these were magnificent machines and were the last barely street-legal production motorcycles.

350 and 500GTL

The parallel twin was Ducati's first completely new production engine design for more than fifteen years, but was inspired by Taglioni's 1965 360-degree prototype overhead valve parallel twin, produced at Berliner's request. When Fabio Taglioni refused to become involved in the new parallel twin's design,

350 AND 500GTL 1975

Engine	Parallel twin four-stroke, air-cooled
Bore (mm)	78 (71.8)
Stroke (mm)	52 (43.2)
Displacement (cc)	496.9 (349.6)
Compression ratio	9.6:1
Valve actuation	Chain-driven single overhead camshaft
Carburetion	Dell'Orto PHF30BS/BD (500) Dell'Orto VHB26FS/FD (350)
Power	35 horsepower at 6,500 rpm (500)
Gears	Five-speed
Front suspension	35mm Marzocchi fork
Rear suspension	Marzocchi 320mm twin shock swingarm
Front brake (mm)	260 dual-disc (single-disc 350)
Rear brake (mm)	160 drum
Tires	3.25x18, 3.50x18
Wheelbase (mm)	1,400
Dry weight (kg)	170 (163)
Top speed (km/h)	170 (147, 350)
Colors	Black frame, blue, green, red
Engine numbers	From DM500.1 500001 (500) From DM350.1 035001 (350)
Frame numbers	From DM500B and DM350B
Production	530 (350GTL) 447 (500GTL)

Ing. Tumidei was given the project and had to start virtually from scratch. With a reduction in manufacturing costs the priority, the parallel twin began life as a compromise and suffered as a result.

While retaining some traditional Ducati features, such as vertically split aluminum crankcases, with the cylinders canted forward 10 degrees, the forged one-piece 180-degree crankshaft was new. Both the main and big-end bearings were plain, the crankshaft without a center bearing. To minimize engine width, the chain camshaft drive was taken from the rear of the clutch hub to a countershaft behind the cylinders. The cylinder head included a more modern 60-degree included valve angle, and unlike the singles and 860GT, ignition was by the traditional battery, coil, and contact breakers.

The single downtube frame followed the usual Ducati practice of using the engine as a stressed member, and the styling mimicked Giugiaro's 860. While the GTL handled and braked well enough, the engine lacked character, the performance was lackluster, and the reliability was suspect. Little loved, it is now considered one of Ducati's least satisfactory models of the 1970s.

The 125 Regolarità was a confused attempt at creating a competitive off-road motorcycle. *Ducati*

125 Regolarità

The third new model to appear for 1975 was the 125 Regolarità, and for a company with little experience in off-road competition, it was an ambitious undertaking. Powering the 125 Regolarità was an all-new two-stroke engine, this featuring a chrome-plated bore and the usual builtup crankshaft running in needle roller and ball bearings. The frame was a double-cradle type, with a large-diameter backbone tube, but although there was a protective plate for the exhaust pipe, the low exhaust limited the Regolarità's off-road ability. And while other manufacturers were including automatic oiling, the Regolarità still required 5 percent premix in the tiny 6-liter gas tank. Produced primarily for Italy and France, the 125 Regolarità was another example of an ill-conceived and uncompetitive machine that characterized this era.

1976

With 860s and parallel twins languishing in showrooms, Franco Zaiubouri replaced De Eccher in September 1975. Zaiubouri immediately revived the relationship with Taglioni, always tense under De Eccher, and set about restoring Ducati's fortunes by facelifting the 860GT and introducing a production 900 Super Sport able to be sold worldwide. But with a large number of unsold stock, only 4,053 motorcycles were built, the lowest annual total yet.

125 REGOLARITÀ SIX DAYS 1975

Engine	Single-cylinder piston-port two-stroke, air-cooled
Bore (mm)	54
Stroke (mm)	54
Displacement (cc)	123.7
Compression ratio	10.5:1
Carburetion	Dell'Orto PHB30
Gears	Six-speed
Front suspension	Telescopic fork
Rear suspension	Twin shock swingarm
Brakes (mm)	Drum 125 (front), 140 (rear)
Tires	3.00x21, 3.75x18 Metzeler 6-days
Wheelbase (mm)	1,420
Dry weight (kg)	108
Top speed (km/h)	120
Engine numbers	From DM125.1 130001
Frame numbers	From DM125C 130001
Production	1,609 (125 Regolarità)
Other Production 1975	250 (239/250M3) 400 (450SCR) 248 (350SCR) 112 (250SCR)

860GTS 1976
Differing from the 860GT and GTE

Engine numbers	From DM860 853000 Approx.
Frame numbers	From DM860S 853000 Approx.
Production	570 (1975)
	510 (1976)

The negative response to the 860GT resulted in the restyled 860GTS for 1976.

860GTS

Responding to poor sales, the 860GT was hurriedly restyled during 1975, becoming the 860GTS. The engine was little changed, but all 860GTS models were electric start. Along with a new rounder 18-liter gas tank and seat, dual-front disc brakes were standard and the handlebar was lower. During 1976, many unsold 860GTs were converted with some GTS equipment, notably the tank and a shorter seat.

750 and 900 Super Sport

It was obvious from the uncompromised specification and limited availability that the 1975 Super Sport was only intended as a one-off series. With the 860 languishing in showrooms, demand for the return of the sporting Ducati led to the Super Sport being put into regular production during 1976. As Berliner wanted the model in the United States, this required a left side gearshift, a steel gas tank, turn signals, and a quieter intake and exhaust system. While the desmodromic engine was as before, the Super Sport gained the 860's ugly Lafranconi mufflers and smaller carburetors with air filters. Contis and 40mm carburetors remained an option and the gearshift was moved to the left via a crossover rod, with the brake on the right while a steel 750 Sport–type gas tank replaced the sculptured fiberglass Imola type.

The 900 Super Sport was adapted for the US market during 1976, now with a steel tank, a left-side gearshift, and turn signals. Conti mufflers and 40mm carbs remained an option.

The 1976 Super Sport was very successful, reestablishing Ducati as a premier manufacturer of sporting motorcycles, and maintaining its reputation for outstanding handling and braking. While some enthusiasts lamented the demise of

750 AND 900 SUPER SPORT 1976
Differing from 1975

Carburetion	Dell'Orto PHF32A
Frame numbers 900SS	From DM860SS 086001
Production	220 (750SS)
	1,020 (900SS)

the unadulterated street race replica, the Desmo SS's soul was left intact despite the introduction of mandated street legality. The blue and silver wire-wheeled Super Sports are now considered another of Ducati's finest production models.

500 Sport Desmo

While the unpopular 350 and 500GTL continued largely unchanged, the 500GTL evolved into the more appealing 500 Sport Desmo during 1976. Leopoldo Tartarini was given the style brief, and Fabio Taglioni provided the desmodromic cylinder head. From the cylinder head down the engine was almost identical to the GTL, but the chassis and styling was new. The frame included a double-front downtube, and the 14-liter fiberglass gas tank was shaped to blend in with the racing-style saddle and ducktail. The wheels were FPS aluminum with a rear disc brake to complement the dual-front discs. Although extremely rare in the United States, the 500 Sport Desmo was one of Ducati's finer styling efforts and a significant improvement over the GTL.

Standard equipment on the 900SS was a frame-mounted half fairing.

1977

Led by the success of the 900SS and new 900SD Darmah, motorcycle production rose to 7,167 during 1977. Incorporating many Japanese components, the Darmah was one of Ducati's most important models yet, while the parallel twin range was expanded to include the GTV.

The 500SD was one of Ducati's finer mid-1970s styling efforts.

500 SPORT DESMO 1976
Differing from the 500GTL

Power	50 horsepower at 8,500 rpm
Front suspension	35mm Paioli fork
Rear suspension	Marzocchi 330mm twin shock swingarm
Rear brake (mm)	260 disc
Dry weight (kg)	185
Top speed (km/h)	185
Colors	Black frame, red or blue with white stripes
Production	441
Other Production 1976	1,177 (125 Regolarità) 400 (350GTL) 285 (500GTL)

ECONOMIC RATIONALIZATION | **127**

Prepared by Phil Schilling, here with the bike, the *Cycle* magazine 750SS was nicknamed "the California Hot Rod."

And despite the failure of the 125 Regolarità, Ducati seemed determined to persevere with the two-stroke Enduro, the Regolarità evolving into the equally uninspiring Six Days. Although official racing involvement was no longer sanctioned, success on the track continued with *Cycle* magazine editor Cook Neilson winning the 50-mile Daytona Superbike race on the Phil Schilling–prepared 750SS. This was Ducati's most significant victory yet in America, Neilson winning easily at average speed of 100.982 miles per hour.

900SD Darmah, 750/900SS, 860GTS

With the failure of the 860GT and GTS to earn a loyal following, Ducati again turned to Leopoldo Tartarini to design a replacement. The resulting 900SD Darmah was a more successful rendition than the Giugiaro 860, but it wasn't only the new style that set it apart. The engine incorporated a number of technical improvements, and Japanese and German electrical components contributed to improved reliability.

Inside the engine were an improved crankshaft, Bosch electronic ignition, and the 900SS's desmodromic cylinder heads. The left-side gearshift was updated, eliminating the sloppy crossover shaft, and the electric-start arrangement was simplified and improved. The Darmah was the first bevel-twin to feature cast-aluminum wheels, Campagnolo on the earliest examples, and also the first with Japanese Nippon Denso instruments and switches. With Tartarini's new sleek bodywork, the Darmah was an extremely successful combination

With the 900SD Darmah, styling became an essential component.

900SD DARMAH 1977
Differing from the 860GT and GTS

Compression ratio	9.3:1
Valve actuation	Bevel gear–driven desmodromic single overhead camshaft
Front suspension	38mm Ceriani or Marzocchi fork
Rear suspension	Marzocchi twin shock swingarm
Front brake (mm)	280 dual-disc
Rear brake (mm)	280 disc
Wheels	Cast 2.15x18, 2.50x18
Tires	3.50x18, 4.25/85x18
Dry weight (kg)	216
Top speed (km/h)	190
Colors	Black frame, red and white
Engine numbers	From DM860 900001
Frame numbers	From DM860SS 900001
Production	1,610

ABOVE LEFT: The distinctive ducktail seat was another product of Tartarini.

ABOVE: The 900SD broke with tradition by including Nippon Denso instruments and a large Bosch headlight.

of style and function and would become one of Ducati's most successful models over the next few years. With most developmental resources directed to the Darmah this year, the existing 750/900SS and 860GTS continued largely unchanged.

350/500SD, 350/500GTV

New for 1977 was the 350 Sport Desmo, produced primarily for the Italian market to entice younger riders limited to 350cc. In response to the poor reception of the GTL's 860-like styling, the 350 and 500GTV replaced the 350 and 500GTL during 1977. Ostensibly 350 and 500 Sport Desmos with valve spring engines, they featured the Sport double downtube frame, Paioli fork, cast-alloy wheels, and rear disc brake. Although the GTV was an improvement over the GTL, no one really wanted a touring parallel twin. The 500SD continued unchanged this year, as did the 500GTL.

BELOW LEFT: Introduced primarily for the Italian market during 1977 was the 350 Sport Desmo.

350SD, 350/500GTV 1977
Differing from the 500SD and 350/500GTL

Dry weight (kg)	18 (500GTV) 181 (350)
Top speed (km/h)	175 (500GTV) 145 (350GTV)
Colors	Yellow with black stripes, blue with black stripes (350SD) Blue (500GTV) Green or black and gold (350GTV)
Frame numbers	From DM500C 550001 (500GTV) From DM350C 035001 (350GTV)
Production	456 (350SD) 1,184 (500SD) 360 (350GTV) 453 (500GTV) 373 (500GTL)

Styled to replicate the Darmah, the 350GTV was intended as a replacement for the 350GTL.

125 Six Days and Regolarità

Although an improved model, the 125 Six Days of 1977 was even less successful than the Regolarità. New for the Six Days was an 8-liter aluminum gas tank, with vented filler cap and a proper enduro high exhaust exiting on the left. The twin Marzocchi shock absorbers allowed for four different mounting positions on the frame and swingarm. While the redesigned frame was stronger, the suspension was still obsolete, and despite more power and less weight, the 125 Six Days was a failure. Another of the more unmemorable Ducatis, with the death of the Six Days in 1979, so ended the era of the Ducati two-stroke.

1978

Although the 900SS and Darmah provided positive signs for the future, and contributed to most of the motorcycles manufactured this year, Ducati's middleweight lineup continued to lack direction, and production slid to only 4,436 during 1978. This heralded a protracted slump in sales over the next few years as the Borgo Panigale plant endeavored to balance an increase in production with continued development. By 1978, it was evident the parallel twin wasn't the middleweight panacea Ducati hoped for and the bevel twins were no longer the standard-setting Superbikes.

Salvation for Ducati came from an unexpected source, Mike Hailwood's spectacular victory in the 1978 Isle of Man Formula 1 TT. Coming out of retirement at the age of thirty-eight,

While an improvement over the Regolarità, the 125 Six Days was still outclassed by more specialized off-road machinery. This was the final Ducati two-stroke. *Ducati*

125 SIX DAYS 1977
Differing from the Regolarità

Compression ratio	14.5:1
Carburetion	Bing 32
Rear tire	4.00x18 Cross-Enduro
Wheelbase (mm)	1,430
Dry weight (kg)	97
Top speed (km/h)	110
Production	998 (Six Days) 300 (Regolarità)

Hailwood captured his fairytale win on a factory-supplied 900NCR. The win was the public relations tonic Ducati desperately needed. As had happened after Imola six years earlier, Hailwood's victory provided immeasurable kudos and led to the release of the Mike Hailwood Replica, a model that would sustain Ducati until 1985.

ABOVE LEFT: Mike Hailwood rode an NCR900 in the 1978 Formula One TT at the Isle of Man. *Roy Kidney*

ABOVE: The 900NCR engine included a host of special parts.

900NCR

While all the factory endurance racers of the 1970s were one-off specials, the creation of the TT1 World Championship saw Ducati offer the 900NCR Formula 1 machine. This was a cataloged model, produced by Scuderia NCR (Nepoti Caracchi Racing). The Formula 1 900NCR was less exotic than the endurance racers but still featured special narrow crankcases incorporating a spin-on oil filter. Other special components included larger valves, lighter pistons, higher lift desmodromic camshafts, lightened crankshaft, closer ratio six-dog gearbox, and straight-cut primary gears. The chrome-molybdenum Daspa frame was lighter and stronger than the production Verlicchi, and the 900NCR was notable for the spectacular attention to detail and beautiful one-piece fiberglass 24-liter tank and seat unit.

Prepared by Steve Wynne and Pat Slinn of Sports Motorcycles in Manchester, Hailwood's 900NCR was painted in Castrol colors of red, white, and green. Hailwood cruised it to win at an average speed of 108.51 miles per hour. His fastest lap was 110.62 miles per hour. It was an astounding victory and vindication, not only of Hailwood's extraordinary ability, but

BELOW LEFT: Mike Hailwood at the Isle of Man in June 1978, on his way to providing Ducati its first world championship. *Ducati*

FORMULA 1 900NCR 1978
Differing from the 900SS

Compression ratio	10:1
Power	105 horsepower at 8,800 rpm
Wheels	Campagnolo MT2.15x18, MT4.00x18
Tires	Michelin 3.50x18 S41 PZ2 Michelin TV2 4.00/5.60x18
Dry weight (kg)	160
Colors	**Red and silver**

ECONOMIC RATIONALIZATION

also of the brilliant balance of power and handling of the 900NCR. Just like Imola, the Isle of Man success would be a pivotal in the racing history of Ducati. It was Ducati's first world championship, and fittingly Mike Hailwood was the company's first world champion.

750/900 Super Sport, 900SD, 900GTS

While the 1976 and 1977 Super Sports were rather hastily conceived, retaining the ignition and crankshaft of the earlier model and incorporating a sloppy gearshift arrangement, most criticisms were addressed for 1978. This saw the incorporation of the 900SD Bosch ignition and crankshaft and a more efficient left-side gearshift. Both solo and dual seats were specified, and most Super Sports included wire-spoked Borrani wheels. During 1978, a black and gold 900SS, with Speedline magnesium alloy wheels, was built for the UK market, some silver 900SSs also receiving the Speedline wheels. Only a few 750SSs were built this year, the last of an illustrious series, and all with Borrani wire wheels. The popular 900SD Darmah continued much as before but was now available in a striking black and gold, while

Only a few 750 Super Sports were built during 1978. The 750SS models featured a silver fairing with blue stripes, and all had Borrani wheels. The final examples like this also included Brembo Gold Series brake calipers.

750/900SS, 900SD, 900GTS 1978
Differing from 1977

Rear brake (mm)	280 disc (900SS)
Wheels	Cast WM3x18; WM4x18 (Some 900SS)
Rear tire	120/90x18 (900SD)
Colors	Silver frame, silver and blue (900SS) Black frame, black and gold (900SS UK) Black and gold (900SD)
Production	30 (750SS) 1,017 (900SS) 1,691 (900SD) 250 (860GTS) 23 (750 Sport) 40 (750GT) 38 (450D/SCR)

Most 1978 900 Super Sports were similar in specification to 1977 examples. A new clutch and primary drive cover with an additional inspection hole characterize the Bosch ignition.

A very small number of blue and silver 1978 900 Super Sports were fitted with magnesium Speedline wheels.

LEFT: A special black and gold 900 Super Sport was produced for the UK market in 1978. Most had a dual seat.

BELOW: The 900SD Darmah was available in black and gold for 1978.

the 860GTS became the 900GTS. Also produced this year for the Australian distributor Frasers were a small number of 750GTs and Sports, built out of existing spares.

PRODUCTION 414 (350SD/GTV)
133 (500SD/GTV)
600 (Six Days/Regolarità)

350/500SD, 350/500GTV, 125 Six Days, Regolarità

With a significant number of unsold stock, and the prospect of the new Pantah for 1979, Ducati began to wind down production of the parallel twins and two-strokes. These were all basically unchanged for 1978.

1979

As demand for the parallel twins continued to decline, production slipped to another all-time low of 3,463 motorcycles, and the most popular model was now the restyled 900 Super Sport.

LEFT: Similar to the 1979 NCR endurance racer, the 1979 NCR Classic machine included a special right-side engine cover with dry alternator.

BELOW: The clutch was also dry, and the narrow crankcases included a spin-on oil filter, also featured on the 1978 900NCR TT1.

Ducati provided this special 900NCR for the 1979 Classic TT, but Hailwood elected not to ride it. The shape of the fairing was new, and it inspired the shape of the production 900 Replica.

BELOW: The first production series of 900 Mike Hailwood Replica appeared surprisingly quickly, barely a year after the race victory.

Joining the 900 Darmah was a striking 900 Super Sport Darmah, and an initial run of Mike Hailwood Replicas was produced for the UK market. Ducati again provided Mike Hailwood NCR racing machines for the 1979 Isle of Man TT, but he couldn't replicate his victory of the previous year. While the bevel-twin range was receiving some attention, the most significant new model was the 500SL Pantah, with a new engine that would shape Ducati's engine design for the next three decades.

900 Mike Hailwood Replica

Toward the end of 1979, the first series of Mike Hailwood Replicas appeared. These were primarily for the British market, each coming with a certificate. The engine was shared with the 900SS, and all early examples came with 40mm carburetors and Conti mufflers. While the frame was also shared with the 900SS, the replica came with striking red, white, and green fiberglass bodywork, patterned after the 1978 and 1979 900NCR. Early replicas had an SS steel gas tank under a fiberglass cover, a one-piece fairing, and the Darmah's Nippon Denso instrumentation and switches.

900 MIKE HAILWOOD REPLICA 1979
Differing from the 900SS

Carburetion	Dell'Orto PHM40B
Power	80 horsepower at 7,000 rpm
Rear suspension	Marzocchi 330mm twin shock swingarm
Rear brake (mm)	280 disc
Tires	100/90x18, 110/90x18
Wheelbase (mm)	1,510
Dry weight (kg)	205
Top speed (km/h)	220
Colors	Red frame, red, white and green
Engine numbers	From DM860 089390 Approx.
Frame numbers	From DM860SS 900001
Frame numbers	DM900R1 905002–906500 Approx.
Production	300

900 Super Sport

The 900SS engine continued largely unchanged for 1979, with Silentium mufflers replacing the Lafranconi, but the Contis were always an option. Fashion dictated a switch from Borrani wire-spoked wheels to magnesium Speedline wheels, with a correspondingly larger rear disc brake. Later in 1979, aluminum FPS wheels replaced the Speedline. Most of these black and gold 900SSs came with a dual seat, but the solo saddle was always optional. Although the cast wheels weren't functionally superior, the black and gold Super Sport was another styling triumph.

900SSD Darmah, 900SD, 900GTS

During 1978, the 900SD Darmah evolved into the 900 Super Sport Darmah, an attractive adaptation of the Super Sport café racer to the Darmah engine and chassis. The engine was shared with the 900SD, early examples with Lafranconi mufflers. Setting the 900SSD apart from both the 900SD and 900SS was the distinctive two-tone blue paintwork. Initially the wheels were magnesium Speedline, aluminum FPS replacing these during 1979. While undoubtedly one of the most attractive Ducatis of the era, it was too heavy to be a real Super Sport, and the overall performance was inferior to the 900SS. The 900SD continued with a few updates for 1979, notably a new seat, and the 900GTS gained the updated Darmah engine, with Bosch ignition, but still with valve spring cylinder heads.

BELOW: Another successful styling effort was the 900SSD. This combined the electric start Darmah with Super Sport ergonomics.

BELOW RIGHT: The final 900GTS of 1979 included many Darmah engine components.

The 900 Super Sport was restyled in attractive black and gold colors for 1979, these with different decals to the few UK 1978 versions.

900SS 1979
Differing from the 1978 and the 900MHR

Colors	Black frame, black and gold
Production	934

900SSD, 900SD DARMAH, 900GTS
Differing from the 1978 900SD

Top speed (km/h)	205 (SSD)
Colors	Two-tone blue (SSD)
Frame Numbers	From DM860SS 903000 Approx. (900SSD) From DM900SD 950001 (900SD)
Production	609 (900SSD 1978 and 1979) 827 (900SD) 150 (900GTS)

ECONOMIC RATIONALIZATION

500SL Pantah

With the parallel twin struggling to find acceptance, during 1976 Fabio Taglioni and Gianluigi Mengoli began designing a replacement. The resulting 90-degree V-twin Pantah was a synthesis of the earlier 500cc bevel-drive Grand Prix engine and belt-drive double overhead camshaft four-valve Armaroli racer of 1973. In 1973, belt camshaft drive was unusual for a motorcycle, but for Taglioni it seemed the ideal alternative to bevel gears as it offered noise reduction and more economical manufacture. The resulting Pantah shared the bore and stroke of the 500cc racer, but incorporated toothed rubber belts to drive the single overhead camshafts. This was such an advanced design that it still forms the basis of some current production Ducatis.

While the family connection was strong, the Pantah differed in a number of details to the bevel-drive twins. Retaining vertically split crankcases, the swingarm pivoted on bearings within the gearbox casing, bringing the pivot as close as possible to the countershaft sprocket. The desmodromic cylinder heads included a more modern 60-degree included valve angle, and the forged one-piece crankshaft featured plain bearing big ends. Supporting this engine was a trellis-type frame, with the engine hanging underneath. The Nippon Denso instruments and instrument panel was all similar to the Darmah SSD, and the half fairing, clip-on handlebars, and rear-set footpegs accentuated the café racer image. Franco Bilancioni provided the Pantah styling in house.

500SL 1979

Engine	90-degree V-twin four-stroke, air-cooled
Bore (mm)	74
Stroke (mm)	58
Displacement (cc)	499
Compression ratio	9.5:1
Valve actuation	Belt-driven desmodromic overhead camshaft
Carburetion	Dell'Orto PHF36A
Power	45 horsepower at 9,050 rpm
Gears	Five-speed
Front suspension	35mm Marzocchi fork
Rear suspension	310mm Marzocchi twin shock swingarm
Front brake (mm)	260 dual-disc
Rear brake (mm)	260 disc
Wheels	2.15x18, 2.50x18
Tires	3.25x18 and 3.50x18
Wheelbase (mm)	1,450
Dry weight (kg)	180
Top speed (km/h)	196
Colors	Black frame, red and silver
Engine numbers	From DM500L.l 600001
Frame numbers	From DM500SL 660001
Production 1979	163 (500SL) 100 (125 Regolarità) 150 (125 Six Days) 300 (350GTV/SD) 130 (500SD)

The first edition 500SL was only built in limited numbers, but it established a new family. *Ducati*

Mallol and Tejedo's modified 900SS that won the 1980 Barcelona 24-hour race.

1980

As two significant models were introduced in limited numbers during 1979 (the 900MHR and 500SL), production increased slightly during 1980, to 4,452. This was still quite a difficult time at Borgo Panigale, and with profitability still questionable during 1980, Ducati moved from EFIM to become part of Finmeccanica in the VM Group. VM was more interested in manufacturing diesels and considered motorcycle production secondary, so the next few years were to be even more problematic. But despite a reduced factory racing effort, Ducati still managed some impressive results, Jose Mallol and Alejandro Tejedo winning the Barcelona 24-hour race at Montjuïc on a modified 900SS. Providing Ducati

The 1980 lineup of factory racing Pantahs. These were very successful in the Junior Italian TT2 Championship. *Ducati*

its first victory in the event since 1975, they covered 757 laps at 74.12 miles per hour. For the 1980 Junior Italian TT2 Championship, Franco Farnè and the racing department prepared four race-kitted Pantahs. With bodywork patterned on the 900NCR, these machines were incredibly successful, and in the hands of Vanes Francini, Pietro Menchini, and Guido Del Piano, they won twelve Junior TT2 races. Del Piano took the championship and this success prompted Fabio Taglioni to release the magnificent TT2 for 1981.

900 MHR, 900SS, 900SSD, 900SD

New for the next series of 900MHR was a 24-liter steel tank and a more easily convertible dual seat, but the MHR was otherwise unchanged. The 900SS, 900SD, and 900SSD Darmah were also little changed but now had FPS wheels and the 900SD no longer had a kick-start.

500SL Pantah

For 1980, the Pantah received new colors and a number of internal engine and gearbox updates but was otherwise unchanged. Although it immediately set a new standard in the 500cc category, restrictive air filters and extremely quiet Conti mufflers limited the 500 Pantah's ultimate performance. To those used to the torque of the larger twins, the Pantah felt anemic.

ECONOMIC RATIONALIZATION | 137

The 1980 900SSD. New this year were FPS wheels and lower footpegs.

The 900MHR received a new tank and seat for 1980, but the style was unchanged.

900MHR, SS, SSD, SD 1980
Differing from 1979

Frame Numbers	From DM900SS 901000 (900MHR)
	From DM900SD 950200 (900SSD)
Production	447 (900MHR)
	753 (900SS)
	705 (900SSD)
	470 (900SD)

The 1980 900SS was very similar to the 1979 version. Most had a dual seat like this.

Still with the revised seat, the 1980 900SD was similar to 1979's model.

500SL 1980 Differing from 1979	
Colors	Black frame, blue and silver
Production 1979	1,459 (500SL)
	424 (350GTV/SD)
	193 (500GTV/SD)

Somewhat surprisingly, given their unpopularity, the existing 350 and 500SD and GTV continued in production this year, but these were primarily for the Italian market.

ABOVE LEFT: Now in new colors, the 500SL went into series production for 1980. *Ducati*

1981

Despite a section of the Borgo Panigale plant diverted from motorcycle manufacture to that of diesel engines, motorcycle production increased to 6,838 for 1981. The most popular models were the 900MHR and new 600SL Pantah, and this year also saw the introduction of the magnificent 600TT2, Massimo Broccoli winning Junior Italian TT F2 Championship on it. Sports Motorcycles also decided to switch from the bevel 900 to a modified Pantah for 1981 TT, Tony Rutter winning the F2 TT at an average speed of 101.91 miles per hour. With only two rounds in the TT World Championship, Ducati then sent Franco Farnè and two TT2 factory machines for Rutter at Ulster, Rutter finishing second in the wet conditions to secure the 1981 World TT Formula 2 Championship.

Massimo Broccoli won the 1981 Junior Italian TT F2 Championship on the new 600TT2. *Ducati*

900 MHR, 900SS, 900SSD, 900SD

Most of the updates to the 900 Replica for 1981 were practical or cosmetic, the main update a two-piece fairing with a removable lower section, a new seat, and side covers to hide the battery and rear carburetor. While the 900SD was largely unchanged this year, the frame was now shared with the SSD. After eight years with little development, the 900SS was mildly updated for 1981, the most significant change a three-dog gearbox replacing the problematic six-dog type. Other changes for 1981 were to the bodywork and a new tank, fairing, seat, and side cover colors and decals.

The 900 Mike Hailwood Replica received a new fairing and side covers for 1981.

ECONOMIC RATIONALIZATION

900MHR, SS, SSD, SD 1981
Differing from 1980

Power	63 horsepower at 7,400 rpm (900MHR)
Frame Numbers	From DM900R 901500 Approx. (900MHR)
Colors	Silver (900SS)
Production	1,500 (900MHR) 1,165 (900SS) 126 (900SSD) 683 (900SD)

Also new for 1981 was a restyled 900 Super Sport.

600SL, 500SL Pantah

For 1981, the 600SL joined the 500SL. Ostensibly identical to the 500, the 600SL included a hydraulic clutch and larger brake calipers, and both models received a new fairing this year. The larger engine provided slightly more torque and went some way to addressing the complaints regarding the 500's midrange performance. Also continuing alongside the Pantahs was limited parallel twin production.

600SL 1981
Differing from the 500SL

Bore (mm)	80
Displacement (cc)	583
Compression ratio	10.4:1
Power	61 horsepower at 9,100 rpm
Rear tire	4.00x18
Dry weight (kg)	187
Top speed (km/h)	201
Colors	Black frame, silver (600SL) White (500SL)
Engine numbers	From DM600L 700001
Frame numbers	From DM600SL 700001
Production	1,095 (600SL) 1,804 (500SL) 172 (500SD) 294 (350/500GTV)

Apart from the new fairing and colors, the 600SL Pantah was similar to the 500SL.

1982

New models this year included the 900S2 and a controversial touring Pantah, the 600TL. Motorcycle production decreased slightly, to 5,665. This was a very successful year for the TT2. Walter Cussigh won all five rounds of the Junior Italian Formula 2 Championship to take the title and Tony Rutter dominated the World TT Formula 2 Championship. Rutter won the Isle of Man TT F2 race at an average speed of 108.50 miles per hour, following this with victories at Vila Real in Portugal and the final round at Dunrod in Ulster. An official factory 750 also appeared at the Bol d'Or d'Italia 24-hour endurance race at Imola in May, a precursor to the TT1 for 1983.

Tony Rutter took his second world championship in 1982 on the TT2.

600TT2

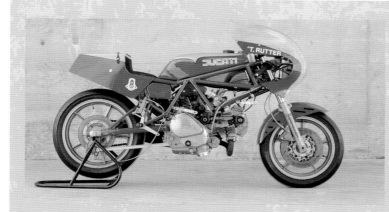

The TT2 was offered as a catalog racer for 1982 and 1983, the 1983 version here with a 16-inch front wheel. *Roy Kidney*

Continuing a tradition of production high specification customer racers, limited numbers of the 600TT2 were available in 1982 and 1983. These spectacular machines were similar to Tony Rutter's four-time World TT2 Championship-winning racer, the production version also featuring a factory-prepared engine with polished crankshaft and con rods. The compact Verlicchi frame with cantilever rear suspension was also shared with the factory bikes. Epitomizing Taglioni's philosophy of maximum performance through light weight and simplicity, the diminutive TT2 was one of the finest of all catalog Ducatis. There was nothing superfluous on the TT2, function determining the form of every component.

600TT2 1982–1983

Bore (mm)	81
Stroke (mm)	58
Displacement (cc)	597
Compression ratio	10.2:1
Carburetion	Dell'Orto PHF36A (1982) Dell'Orto-Malossi 41mm (1983)
Power	76 horsepower at 10,500 rpm (1982) 78 horsepower at 10,500 rpm (1983)
Gears	Five-speed
Front suspension	35mm Marzocchi magnesium fork
Rear suspension	Marzocchi PVS 1 cantilever swingarm
Wheels	2.15x18, 2.50x18 (1982) 3.50x16, 3.50x18 (1983)
Front brake (mm)	280 dual-disc
Rear brake (mm)	260 disc
Dry weight (kg)	130
Colors	Red frame, red and yellow
Production	20 (1982) 30 (1983)

900S2, SD 1982
Differing from the 900SS

Wheelbase (mm)	1,500 (900S2)
Dry weight (kg)	190 (900S2)
Top speed (km/h)	205 (900S2)
Colors	Black frame, bronze (900S2) Blue, burgundy, gray (900SD)
Engine numbers (900S2 kick-start)	From DM860 095000 Approx.
Engine numbers (900S2 electric start)	From DM860 907000 Approx.
Frame numbers 900S2	From DM860SS 092001 Approx.
Production	649 (900S2) 317 (900SD) 1,549 (900MHR) 335 (900SS)

TOP RIGHT: The 900S2 was an attempt to combine the finest attributes of the Super Sport and the Darmah.

RIGHT: New colors set the 1982 Darmah apart, but it was still very similar to previous versions.

BELOW RIGHT: The 600TL was afflicted with unfortunate styling and didn't sell well.

900S2, 900MHR, 900SS, 900SD

The 900SS continued for one more year, but during 1982 the 900S2 gradually replaced it. While the S2 was primarily a styling exercise, it was also an attempt to widen the Super Sport's appeal by introducing an electric start. The S2 frame was an amalgam of the Super Sport and Darmah, the fairing was from the 600SL Pantah, with the electrical system, switches, and instruments of the 900MHR. The 900MHR continued as before, and the final 900SD Darmah included new colors and graphics.

600TL 1982
Differing from the 600SL

Power	58 horsepower at 8,700 rpm
Dry weight (kg)	177
Top speed (km/h)	183
Colors	Black frame, white
Engine numbers	From DM 600L 650001
Frame numbers	From DM600TL 060001
Production	288 (600TL) 1,350 (600SL) 408 (500SL)

600TL, 600SL, 500SL Pantah

Occasionally Ducati releases a model that is a triumph of styling over function, and one of these was the 600TL. The styling was extreme, but underneath was a reliable and competent motorcycle. The basic engine and chassis was shared with the sporting 600SL, but special features included a rectangular Carello headlight, a small handlebar fairing, and smaller front brake calipers. During 1982, the 600SL received a new cable-operated clutch and Mike Hailwood Replica colors.

The 600SL was available in MHR colors for 1982.

350XL

Introduced specifically for the Italian market to accommodate a tax break for motorcycles below 350cc, the 350 XL was a parts-bin special, essentially a downsized 500SL with a 600TL handlebar fairing and higher handlebars.

1983

By 1983, the situation at Borgo Panigale was grim. Although 3,909 motorcycles were manufactured this year, the VM Group sought to end motorcycle manufacture and began negotiation with Cagiva in Varese to supply engines for Cagiva motorcycles. As this was a slow and protracted process, the existing range continued much as before, with a new electric-start 900MHR introduced later in the year. Tony Rutter again won the TT2 World Championship, the successful TT2 also growing into a 750cc TT1 this year. Rutter finished third in the Battle of the Twins race at Daytona on a 750. Then at Montjuïc Park in July, a factory TT1 won the nonchampionship 24-hour race in the hands of Benjamin Grau, Enrique de Juan, and Luis Reyes.

900MHR, 900S2

This year an electric-start 900MHR joined the existing kick-start version, the electric-start motor including new outer engine covers and a dry clutch. The frame was now shared with the 900S2, accompanied by new side covers and a taller fairing. Other updates included Oscam wheels that could accept tubeless tires. The 900S2 was offered this year with a belly pan, another dubious triumph of styling over function.

A parts bin model created primarily for the Italian market, the 350XL was more attractive than the TL.

350XL
Differing from the 500SL

Bore (mm)	66
Stroke (mm)	51
Displacement (cc)	349
Compression ratio	10.3:1
Carburetion	Dell'Orto PHF30 A
Power	40 horsepower at 9,600 rpm
Dry weight (kg)	177
Top speed (km/h)	170
Colors	Black frame, red and black
Engine numbers	From DM350L 030001
Frame numbers	From DM350XL 300001
Production	749

ABOVE: Tony Rutter won his third TT2 World Championship in 1983.

RIGHT: The electric-start 900MHR received new outer engine cases.

BELOW RIGHT: The 900S2 continued with the earlier engine.

650SL, 600SL, 500SL, 600TL Pantah

The final sporting Pantah before the advent of the 750F1 was also arguably the best, and certainly one of the rarest. Created to homologate a longer stroke for the racing 750TT1, the 650SL also received a moderate bore increase. The chassis and bodywork was essentially identical to the 1983 600SL, now with Oscam wheels and large rectangular CEV indicators and taillight. Specific for the 650 was a new instrument panel and instruments. Although short-lived, the 650SL was the finest Pantah, the 650 engine also living on in the Alazzurra. Like the 600SL, the 600TL also received Oscam wheels this year, but the 500SL was unchanged for its final year.

350SL, 350TL, 350XL

Introduced alongside the 350XL for 1983 were the 350TL and 350SL. The 350TL was ostensibly a smaller 600TL, while the 350SL, other than its silver FPS wheels and smaller brake calipers, looked similar to the red 600SL. This unremarkable trio was little seen outside Italy.

900MHR, S2 1983
Differing from 1982

Compression ratio	9.3:1
Power	72 horsepower at 7,000 rpm
Wheelbase (mm)	1,500 (MHR)
Dry weight (kg)	212.5 (MHR)
Top speed (km/h)	222 (MHR)
Colors	Red frame, black (900S2)
Engine numbers (900MHR electric start)	From DM860 907800
Frame numbers (900MHR electric start)	From DM900R1 905002
Production	1,467 (900MHR) 382 (900S2)

The finest Pantah, and the most striking looking, was the short-lived 650SL.

650SL 1983
Differing from the 600SL

Bore (mm)	82
Stroke (mm)	61.5
Displacement (cc)	649
Compression ratio	10:1
Power	63 horsepower at 8,500 rpm
Dry weight (kg)	180
Colors	Red frame, red and yellow
Engine numbers	From DM650L 610001
Frame numbers	From DM650SL 065001
Production	288 (650SL) 551 (600SL) 513 (600TL)

Another model produced primarily for the Italian market was the 350SL.

350SL, XL, TL 1983
Differing from 1982

Dry weight (kg)	176 (SL)
Top speed (km/h)	180 (SL)
Colors	Red (TL), red frame, red (SL)
Production	677 (XL, TL, SL)

1984

Under the agreement with Cagiva, production of Ducati motorcycles was meant to cease at the end of 1984, but as the negotiations moved slowly, VM provided Ducati a reprieve for another year. Most Pantah engines were destined for Cagiva in Varese and production slipped to only 1,765. Despite the difficulties, somehow Ducati still managed to introduce the next generation bevel twin, the Mille, and also develop the TT1 750. Tony Rutter won his fourth consecutive TT2 World Championship, also campaigning the TT1.

This customer TT1 was ridden to victory in the 1984 Barcelona 24-hour race.

750 TT1 1984
Differing from the TT2

Bore (mm)	88
Stroke (mm)	61.5
Displacement (cc)	748
Power	80 horsepower
Colors	Red frame, red and blue

RIGHT: Only a small number of TT1s were available during 1984.

BELOW: The 1984 factory machines had rising rate rear suspension. This is Walter Villa at the 1984 Bol d'Or at Paul Ricard.

750 TT1

For 1984, a factory 750 kit was available to transform the TT2 into a 750 prior to a limited run of TT1 Replicas becoming available. When the TT1 was finally produced, these were ostensibly the earlier TT2 fitted with a 748cc engine. Based on Tony Rutter's 1984 TT1, the Verlicchi frame was as before, but with a wider aluminum cantilever swingarm to accommodate wider wheels along with an endurance-style quick-change wheel and brake assembly. Although not as exotic as the factory racers, the customer TT1 was still surprisingly effective, Grau, Garriga and Reyes riding number 24 (of twenty-five) to victory in the nonchampionship 1984 Barcelona 24-hour race.

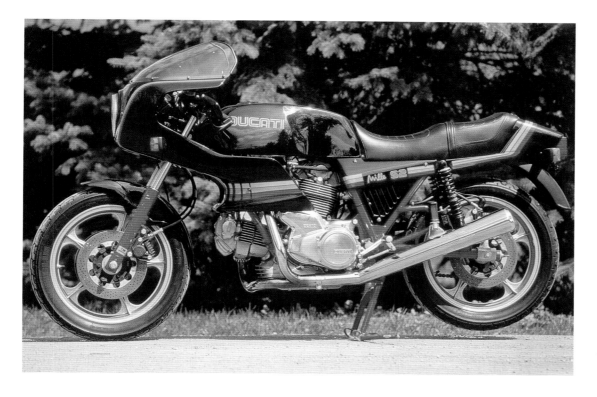

The Mille S2 shared the Mille MHR engine. The outer covers were similar to the final 900MHR.

Mille Mike Hailwood Replica, Mille S2, 900MHR, 900S2

The final bevel-drive twin emerged during 1984, the Mille, arguably the finest bevel engine yet. Shared with the final 900MHR were the dry clutch and external engine covers, but the Mille incorporated many updates, including a larger bore and longer stroke, larger intake valve, and forged one-piece crankshaft with plain big-end bearings. The chassis and bodywork for both the Mille MHR and S2 was almost identical to the 1983 and 1984 900MHR and 900S2. Relatively rare, and potentially the most reliable, the Mille was the definitive development of the classic bevel twin. Alongside the Mille MHR and S2, the 900MHR and S2 continued, much as they were toward the end of 1983. Only a few 600 Pantahs were built this year, and the parallel twin was finally finished for good.

The Mille MHR looked very similar to the 1983–1984 900 electric-start version, but the engine was quite different.

MILLE MHR, MILLE S2
Differing from the 900MHR and S2

Bore (mm)	88
Stroke (mm)	78
Displacement (cc)	973
Power	76 horsepower at 6,700 rpm
Dry weight (kg)	198
Top speed (km/h)	222
Colors	Red frame, red, white and green (MHR) Red frame, black (S2)
Engine numbers	From ZDM1000 100001 From ZDM1000 100700 (S2) Approx.
Frame numbers	From ZDM1000R 100001 From ZDM1000S2 100001 (some early S2s had 1000R frames)
Production	662 (Mille MHR) 71 (Mille S2) 795 (900MHR) 205 (900S2) 32 (600SL)

CHAPTER 6
THE CAGIVA ERA
THE PASO, 851, AND 888, 1986–1993

Marco Lucchinelli with Taglioni and mechanic Pedretti after winning the 1986 Daytona Battle of the Twins race. *Ducati*

Virginio Ferrari won the 1985 Italian F1 Championship on the factory 750.

The transfer of control from the Italian government-owned EFIM Group to Varese-based Cagiva in northern Italy ushered a new era for Ducati. Owned by the Castiglioni brothers, Cagiva immediately initiated a change in direction, with an emphasis on widening Ducati's appeal. Cagiva also endeavored to make production more profitable and instigated a program of new model development that was largely absent in the early 1980s.

1985

During this transitional period, Cagiva undertook the development of the new generation Paso, and in the meantime Ducati's existing lineup of Mille bevel twins continued, albeit on a reduced scale, this year also seeing the introduction of the long-awaited race replica 750F1. But while motorcycle production at Borgo Panigale remained very low this year, with only 1,924 manufactured (including 999 Cagiva Alazzurras built in Varese), Cagiva provided more racing support. This included factory entries in the Endurance World Championship and the Italian F1 series, Virginio Ferrari winning the Italian F1 championship from Marco Lucchinelli and TT1s filling out the top seven places. Alongside the few hundred F1s this year, the Mille MHR and S2 continued unchanged.

750F1

With the factory TT2 continuing to win world championships, it was inevitable that the TT2 would eventually become a production road model. Unfortunately, serious economic problems during the early 1980s saw the production model delayed for several years, but when it was released for 1985 as the 750F1, it was worth the wait. The 750F1 almost came too late, its release coinciding with the sale of Ducati to Cagiva, and one of Cagiva's initial changes was to scale down the production of the traditional Ducati (like the F1) to concentrate on more dubious models (like the Paso). Thus, the F1 represents the end of the pre-Cagiva era.

The black F1 engine was similar to the Pantah, retaining the small valves, with a hydraulic clutch and an oil cooler mounted underneath the headlight. Carburetion was the same as the 650SL, but the F1 frame provided no room for an air filter, and the exhaust was a black-chrome Conti two-into-one. The frame was modeled on the TT2, but was modified to accept timing belt covers, a centerstand, and rear section to support the battery and seat. The aluminum fuel tank was similar to the TT2, with the instruments and electrical equipment from the Mille MHR. Although it didn't offer exceptional engine performance, the 750F1 was a commendably light and compact motorcycle, befitting a machine descended from a pure racer.

RIGHT: The 750F1 finally went into production during 1985, still with the earlier Giugiaro graphics.

BELOW RIGHT: The 750F1 frame was based on the racing TT2.

750F1 1985
Differing from the 650SL

Bore (mm)	88
Displacement (cc)	748
Compression ratio	9.3:1
Front suspension	38mm Marzocchi fork
Rear suspension	Marzocchi PVS4 single-shock cantilever swingarm
Front brake (mm)	280 dual-disc
Wheels	MT2.50x16 and MT 3.00x18-inch
Tires	120/80V16, 130/80V18
Wheelbase (mm)	1,400
Dry weight (kg)	175
Top speed (km/h)	200
Colors	Red frame, red white and green
Engine numbers	7500001–7500594
Frame numbers	ZDM750R 7500001–7500594
Production	625 (750F1) 299 (Mille MHR and S2)

1986

Cagiva immediately indicated its intent on broadening Ducati's appeal, first with the 650 Alazzurra in 1985, followed in 1986 by the Indiana cruiser and the radical Paso. The Paso was Cagiva's first all-new model and represented a significant departure for Ducati. Designed to be more user friendly, it was supposed to be simple to service and a balanced motorcycle. Unfortunately, while the concept was sound, the execution left something to be desired, and as the first Ducati with the engine completely covered, it was almost too radical. But while it may have come under criticism from the traditional Ducati enthusiast, the Paso was an important step forward. It also initiated an increase in production, this rising to 5,979 during 1986.

The Castiglioni brothers were also committed to expanding the racing program and at Daytona Ducati fielded a team of five factory machines, Lucchinelli riding an experimental 851cc TT1 to victory in the Battle of the Twins race. He followed this with a win in the opening round of the World TT Formula 1 Championship at Misano, and a victory in the BOTT race at Laguna Seca. But while this success with the aging Pantah was encouraging, Cagiva saw

The first Desmoquattro was the 748, here at the 1986 Bol d'Or at Paul Ricard. *Ducati*

its future in Superbike racing and was sufficiently farsighted to sanction the development of an engine embracing up-to-date thermodynamic and engine management technology. They engaged Massimo Bordi to design a double overhead camshaft four-valve desmo with Weber Marelli I.A.W. Alfa/N open-loop fully mapped electronic fuel injection. Using modified Pantah crankcases, the 748cc prototype was entered in the Bol d'Or 24-hour endurance race in September 1986. Ridden by Marco Lucchinelli, Juan Garriga, and Virginio Ferrari, the 94-horsepower 748 was in seventh place after thirteen hours before retiring with a broken con-rod bolt.

750 Paso

Designed by one of the greatest modern motorcycle designers, Massimo Tamburini, the name Paso was derived from the Italian racer Renzo Pasolini, who was killed at Monza in 1973. Pasolini had a long association with Aermacchi, a company acquired by Cagiva in 1978. Using the 1986 750F1 engine as a basis, the 750 Paso incorporated a number of significant updates, notably a reversed rear cylinder head to allow the installation of a dual-throat automotive-style Weber carburetor and large single air cleaner. Due to the increased heat generated from the rear exhaust valve, the oil cooling system on the Paso included twin oil radiators, one on either side of the engine in the fairing.

When it came to the chassis and running gear, the Paso was also unlike previous Ducatis. The box-section steel frame was a traditional double downtube, full cradle design,

750 PASO 1986
Differing from the 750F1

Carburetion	Weber 44DCNF
Power	72.5 horsepower at 7,900 rpm
Front suspension	42mm Marzocchi M1R
Rear suspension	Rising rate swingarm
Rear brake (mm)	270 disc
Wheels	3.75x16, 5.00x16
Tires	130/60x16, 160/60x16
Wheelbase (mm)	1,450
Dry weight (kg)	195
Top speed (km/h)	210
Colors	Red, white, blue
Engine numbers	From ZDM750 LP 750001
Frame numbers	From ZDM750 P 750001
Production	1,025

ABOVE: With its full coverage bodywork and solid screen, the 750 Paso was completely unlike any other Ducati and took some getting used to.

BELOW: The 750F1 was slightly updated and improved for 1986.

with a much steeper steering head angle than was normal for a Ducati (25 degrees). The aluminum swingarm no longer pivoted solely in bronze bushes in the engine crankcases, and three linkages operated the single shock absorber. This Soft Damp system was considerably more sophisticated and effective than the F1 monoshock. Integral to the Paso's design were the 16-inch Oscam wheels fitted with radial tires and enclosed fiberglass bodywork with a solid windshield and integral mirrors.

750F1

An updated 750F1 was produced for 1986, this alleviating some of the shortcomings of the 1985 version. The engines received stronger crankcases, crankshaft, and gearbox, along with larger valves and higher lift camshafts. Updated instruments included white-faced Veglias, and the tailpiece was restyled, while the front fork was an improved Forcella Italia (formerly Ceriani). Toward the end of 1986, the final batch of 750F1s was built, bodywork updates including a two-piece Montjuïc front fender, a dual seat with grab handles, and a replaceable solo seat cowling.

750F1 Montjuïc

Although Cagiva was committed to replacing the existing model lineup, it released the limited-edition 750F1 Montjuïc during 1986. This impressive motorcycle was a traditional race replica, supplied with racing tires, no turn signals, and a loud Verlicchi exhaust. Named after one of Ducati's happiest racing hunting grounds, the Montjuïc Park circuit in Barcelona, Spain, the Montjuïc was a development of the

750F1 1986
Differing from 1985

Compression ratio	10:1
Power	75 horsepower at 9,000 rpm
Front suspension	40mm Forcella Italia
Engine numbers	From 7500595
Frame numbers	From ZDM750R 7500595
Production	1,177

The second series 1986 750F1 included a dual seat.

750F1, the engine including new crankcases, vented dry clutch, larger inlet ports, much hotter cams, and larger carburetors.

But for an aluminum swingarm, the frame was the same as the F1 while the wheels were composite Marvic/Akront magnesium/aluminum. With polished aluminum rims and three magnesium spokes, these were much lighter than the Oscams of the F1. The front brakes were significantly upgraded to include racing Brembo Gold series four-piston calipers.

Each Montjuïc came with a numbered plaque on the fuel tank, and they were loud and uncompromising machines. Although they could have been more effective if they were more closely related to the factory TT1 racers of 1984 and 1985 (with a rising rate rear suspension), they remain beautiful and highly desirable. In every respect, the Montjuïc was a considerably faster and more effective sports motorcycle than a stock 750F1.

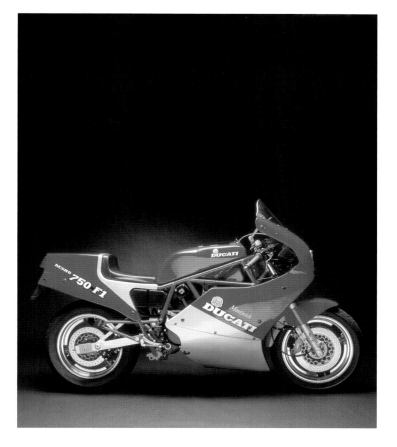

One of the most spectacular production Ducatis of the 1980s: the 750F1 Montjuïc.

750F1 MONTJUÏC 1986
Differing from the 750F1

Carburetion	Dell'Orto PHM40N
Power	95 horsepower at 10,000 rpm
Front suspension	40mm Forcella Italia
Wheels	3.50x16, 4.25x16
Tires	12/60-16, 18/67-16
Dry weight (kg)	155
Colors	Red frame, red and silver
Engine numbers	From 7500595
Frame numbers	ZDM750M 001–200
Production	200

350/400F3 1986
Differing from the 750F1

Bore (mm)	66 (70.5)
Stroke (mm)	51
Displacement (cc)	349 (398)
Compression ratio	10.3:1 (10.4:1)
Carburetion	Dell'Orto PHF30M
Power	42 horsepower at 9,700 rpm
Front suspension	35mm Marzocchi fork
Front brake (mm)	260 dual-disc
Tires	100/90x18, 120/80x18
Dry weight (kg)	180
Top speed (km/h)	175
Colors	Red frame, red and white
Engine and frame numbers	From DM350L 034001 From DM350R 350001
Production	976

The 350F3 was a downsized 750F1, primarily for the Italian market. *Ducati*

350/400F3

A 350 and 400F3 was also built during 1986, primarily for the Italian and Japanese markets. The engine was similar in specification to the final 350SL, while the frame was shared with the 750F1. The biggest difference between the 750 and its smaller brothers was in the brakes and suspension, the 350/400 with a nonadjustable Marzocchi fork and smaller front brakes.

350, 650, 750 Indiana

While they did much to preserve Ducati's heritage and future, Cagiva was also responsible for several dubious models, in particular the Indiana cruiser. As with the earlier Americano, the Indiana was an endeavor to win the US market but was totally misguided in conception and application, tarred by the extremely poor quality finish. Three engine sizes were available, most sharing the basic 650cc engine of the Alazzurra, but with a wider ratio five-speed gearbox and Bing constant-vacuum carburetors. The 350 was built for the Italian market, and the 750 shared the basic engine with the Paso. The Indiana's external alloy engine covers, including the clutch cover and timing belt covers, were chrome plated and especially subject to pitting from corrosion.

Despite its usual Ducati rear twin shock absorber setup, the full cradle square-section steel frame was similar to the Paso, while the cast-aluminum swingarm was box section, pivoting only in the engine cases. Also like the Paso, the Indiana engine featured a reversed rear cylinder head. The Indiana was short-lived, and it wasn't one of Ducati's memorable models. Cagiva was responsible for some of the greatest ever Ducatis but also some of the most horrendous, the Indiana falling firmly in the latter category.

1987

A priority in 1986 was the introduction of the 750 Paso and Indiana, and these continued with only minor updates for 1987. Unfortunately, neither was as successful as anticipated, and production declined to 4,317 during 1987. While the 750F1 was now finished, a new Limited-Edition Laguna Seca series was offered, but ostensibly this year was developmental, with several new models slated for 1988 release.

ABOVE: Marco Lucchinelli took the prototype 851 to an easy victory in the 1987 Daytona BOTT race. *Ducati*

BELOW: A plumber's nightmare, the first 851 was complicated but effective.

Following its brief showing in September 1986, the new racing four-valve twin grew to 851cc for 1987. New crankcases allowed for a six-speed gearbox and continued development resulted in the power increasing to 120 horsepower at 11,500 rpm. In March 1987, the 851 was entered in the Daytona the Battle of the Twins race, Lucchinelli winning easily and timed at an impressive 165.44 miles per hour.

350, 650, 750 INDIANA 1986
Differing from the Pantah

Bore (mm)	66 (350) 82 (650) 88 (750)
Stroke (mm)	51 (350) 61.5 (650 and 750)
Displacement (cc)	649, 349, 748
Compression ratio	10:1
Carburetion	Dell'Orto PHF30DD (350) Bing 64-32 (650, 750)
Power	38 horsepower at 9,250 rpm (350) 53 horsepower at 7,000 rpm (650) 53.6 horsepower at 7,000 rpm (750)
Front suspension	40mm Marzocchi
Front brake (mm)	260 disc
Rear brake (mm)	280 disc
Wheels	2.50x18, 3.50x15
Tires	110/90x18, 140/90x15.
Wheelbase (mm)	1,530
Dry weight (kg)	180
Colors	Black frame, black, red, purple (350) Black, blue (650) Black, silver (750)
Frame numbers	From ZDM350C 350001 (350) From ZDM650C 650001 (650) From ZDM750C 750001 (750)
Production 1986	1,814 (Indiana) 250 (Mille MHR)

750F1 LAGUNA SECA 1987
Differing from the Montjuïc

Rear brake (mm)	270 (Laguna Seca)
Wheels	3.75x16, 5.00x16
Tires	130/60x16, 160/60x16
Frame numbers	From ZDM750LS 750001
Production	296 (Laguna Seca)
Other production 1987	94 (400F3)
	438 (Indiana)
	3,278 (750 Paso)

ABOVE: The second series of Limited-Edition 750F1 was the Laguna Seca.

ABOVE: Although unreliable, Lucchinelli's 1988 factory 888 was extremely fast and could have won the World Superbike Championship if Ducati had contested all the rounds.

BELOW: Marco Lucchinelli on the 888 at Sugo, Japan, in 1988. He could only manage twelfth place. *Ducati*

750F1 Laguna Seca

Marco Lucchinelli's success in the Battle of the Twins race at Laguna Seca in 1986 prompted Ducati to name the next series of Limited-Edition 750F1s the Laguna Seca. Each Laguna Seca also came with a Marco Lucchinelli decal autograph on the gas tank. Ostensibly the engine was identical to the Montjuïc, but to reduce noise, the muffler included a larger canister and riveted aluminum cover. US versions received a different muffler, a Conti similar to that on the 1986 750F1.

The frame was similar to the Montjuïc, but the Oscam wheels and brake discs were shared with the 750 Paso. The front brake calipers were still the Montjuïc four-piston Brembo racing, but the Paso wheels and discs undoubtedly detracted from the purity of the Montjuïc, compromising performance and handling. While most Laguna Secas came with a solo seat, some were also produced with a dual seat.

1988

With the release of the production 851 Desmoquattro, Ducati entered the modern era. No longer burdened by obsolete design parameters, the Desmoquattro was at the forefront of modern technology, and this resulted in an era of racetrack dominance. Before the 851 Ducatis, significant race wins were sporadic, but once the Desmoquattro's gremlins were expunged, it became Ducati's most successful racer yet. Alongside the new 851 was a larger, water-cooled Paso and *nuovo* 750 Sport, while the few unsold US 750F1s received a new exhaust system so they were street legal, the distributor (Cagiva/North America, Glendale, CA) terming these the 750F1S. Also built this year was a series of six-speed 400F3s for the Japanese market, but with the 750 Paso not proving as popular as expected, motorcycle production slipped to 3,950 this year.

With the establishment of the World Superbike Championship, Ducati was offered the perfect platform for its new 851, and Lucchinelli won the very first World Superbike race at Donington. Unfortunately, electrical and crankshaft problems resulted in a number of retirements, but by the end of the season, the 162-kilogram 851 (now displacing 888cc) produced 122 horsepower at 11,000 rpm.

The first production Desmoquattro was the 1988 851 Strada. The mirrors were incorporated in the fairing like the Paso, and the wheels were 16-inch.

The 851 Superbike Kit was built to homologate the 851 for Superbike Racing. Standard equipment included racing wheels and brakes and a headlight and taillight. *Ducati*

851 Strada and Superbike Kit 1988

Only a small number of 851s was produced during 1988, and these were the 851 Strada and 851 Superbike Kit. The Superbike Kit was a homologation model for World Superbike racing. The motor on both was quite similar, with the same Weber fuel injection system with twin injectors per cylinder. The 851s were considerably more powerful than the air-cooled 750 and 900, but the output on these early models was still quite conservative.

The tubular-steel frame was derived from the F1, but with an aluminum swingarm and a rising rate linkage rear suspension setup similar to the Paso. The wheels on the Strada were the 16-inch Marvic/Akront composite wheels previously fitted to the Montjuïc, while those of the Superbike Kit were Marvic 17-inch racing magnesium. The wheels were shod with Michelin racing slicks, yet the Superbike Kit still came with an electric start and a headlight and taillight.

As they were limited production models, many components were individually crafted on the first series, including the 20-liter aluminum fuel tank and rear suspension rocker linkage. The design also followed Ducati's usual evolutionary style and was the result of collaboration between Franco Bilancioni and Franco Farnè. Unfortunately, while the quality of execution was high, the 1988 851 Strada was a flawed motorcycle as the 16-inch wheels provided idiosyncratic steering and handling. The Superbike Kit was also too heavy and slow to be a competitive Superbike racer.

851 STRADA, SUPERBIKE KIT 1988

Engine	Twin-cylinder four-stroke, liquid-cooled
Bore (mm)	92
Stroke (mm)	64
Displacement (cc)	851
Compression ratio	10.2:1 (10.6:1, Kit)
Valve actuation	Belt-driven desmodromic double overhead camshafts, four valves per cylinder
Injection	Weber I.A.W. CPU P7
Power	102 horsepower at 9,000 rpm 120 horsepower at 10,000 rpm (Kit)
Gears	Six-speed
Front suspension	42mm Marzocchi M1R
Rear suspension	Marzocchi rising rate swingarm
Front brake (mm)	280 dual-disc
Rear brake (mm)	260 disc
Tires	130/60x16, 160/60x16 12/60x17, 18/67x17 (Kit)
Wheelbase (mm)	1,460
Dry weight (kg)	185 (165, Kit)
Colors	Silver frame, red, white, green
Production	518

All Santamonicas were dual seat, and the wheels and brakes were shared with the Montjuïc rather than the Laguna Seca.

750 SANTAMONICA 1988
Differing from the Montjuïc and Laguna Seca

Colors	Red frame, red and white
Frame numbers	From ZDM750LS 750303
Production	204 (Santamonica 1987)
	300 (400F3)

906 PASO 1988
Differing from the 750 Paso

Engine	Twin-cylinder four-stroke, liquid-cooled
Bore (mm)	92
Stroke (mm)	68
Displacement (cc)	904
Compression ratio	9.2:1
Power	88 horsepower at 8,000 rpm
Gears	Six-speed
Dry weight (kg)	205
Top speed (km/h)	225
Colors	Red, blue, black
Engine numbers	From ZDM904SC 000001
Frame numbers	From ZDM906 PC 000001
Production	1,302 (906)
	560 (750 Paso)

750F1 Santamonica

Built toward the end of 1987 but sold as 1988 models, the final series of Limited-Edition 750F1s was the Santamonica. Named after the Misano Autodromo Santamonica, where Lucchinelli won the opening round of the World TT Formula 1 Championship in 1986, the Santamonica was built primarily for the Japanese market, at that time Ducati's most important one. The engine was identical to the Laguna Seca, but the wheels were the Marvic/Akront of the Montjuïc and 851 Strada. All Santamonicas were dual seat, and all three 750F1 limited edition series were among of the finest production Ducatis of the 1980s.

Evolution of the 750 Paso continued with the liquid-cooled 906 Paso. The chassis was much as before.

906 Paso

The evolution of the 750 Paso continued with the 906. The liquid-cooled 906 engine was now based on the new 851 large crankcase engine, but with two valves per cylinder. The "6" in 906 also indicated a six-speed gearbox. Other updates over the 750 included larger valves, a stronger crankshaft, new timing belt covers, and Digiplex Magneti Marelli digital electronic ignition. The frame and 16-inch wheels were shared with the 750 Paso, but the front fork was longer. Like the 750 Paso, the 906 should have been successful, but had disappointing sales numbers. Not only was the styling radical, the low-profile tires and 16-inch wheels provided unusual steering and handling and the Weber carburetor was problematic. The Pasos were idiosyncratic motorcycles that haven't stood the test of time, but Tamburini made amends with his next effort, the 916.

750 Sport

With the 750F1 phased out, and the 750 Paso becoming the sport touring model in Cagiva's new Ducati lineup, the 750 Sport was introduced to complement the expensive 851. Named to create an association with the legendary round-case bevel-drive 750 Sport, unfortunately the *nuovo* 750 Sport suffered from insufficient development and was beset with problems throughout its production life.

The 750 Sport was another amalgam of existing models, with the 750 Paso Weber-carbureted engine, 16-inch wheels, and brakes in a modified 750F1 steel-trellis frame with a Verlicchi aluminum swingarm and budget suspension. Although more widely accepted than the 750 Paso, the 750 Sport was seen as a budget parts bin model and a poor substitute for the 750F1. Flawed and ugly, this wasn't one of Ducati's better efforts.

1989

The 851 Strada went into regular production during 1989, and it was the only new model this year. The existing 906 Paso and 750 Sport continued largely unchanged, and although 5,060 motorcycles were built in 1989, a significant proportion

BELOW LEFT: The *nuovo* 750 Sport was a flawed replacement for the 750F1.

750 SPORT 1988
Differing from the 750 Paso

Power	72 horsepower at 8,500 rpm
Front suspension	40mm Marzocchi
Rear suspension	Marzocchi cantilever swingarm
Wheelbase (mm)	1,450
Dry weight (kg)	195
Top speed (km/h)	210
Colors	Red frame, red and blue, red and silver, black and silver (1989)
Engine numbers	From ZDM750LS 750001
Frame numbers	From ZDM750S 750001
Production 1988	1,242 (750 Sport) 36 (650 Indiana)

Now in new colors and with 17-inch wheels, the 851 Strada went into regular production during 1989.

was produced after September for the 1990 model year. Ducati continued to mount a serious challenge for the World Superbike Championship with Raymond Roche signed as the No. 1 rider this year. The 888's power was up to 128 horsepower and the weight down to 158 kilograms. Unfortunately, mechanical and electrical problems resulted in several nonfinishes, and although Roche won five races, he only was third in the championship points. This year, the factory also supplied a limited number of 851 Racing, or Lucchinelli, Replicas.

851 Strada

No longer a limited production model, and showing evidence of some cost cutting, the 851 Strada was significantly improved for 1989. The engine included a single fuel injector per cylinder, and Brembo 17-inch wheels replaced the earlier 16-inch; the front brake disc diameter increased.

1990

Ducati's perseverance with the sophisticated 888 World Superbike racer paid off during 1990; Raymond Roche won Ducati's first World Superbike Championship. The 851's power was increased to 130 horsepower at 11,000 rpm, but despite problems with cracking crankcases, Roche managed eight victories on the 147-kilogram factory Superbike. Ducati also made twenty Roche Replica 851 customer racers available during 1990.

By 1990, it was evident some of Cagiva's new models were unsuccessful, and with the restructuring of the lineup, production slipped to 4,603. While the 851 Strada continued largely unchanged except for a dual seat, a higher performance Sport Production version was available, and

851 STRADA 1989
Differing from the 1988 851 Strada

Compression ratio	11:1
Power	105 horsepower at 9,000 rpm
Front brake (mm)	320 dual-disc
Rear brake (mm)	245 disc
Wheels	3.50x17, 5.50x17
Tires	120/70x17, 180/55x17
Wheelbase (mm)	1,430
Dry weight (kg)	190
Top speed (km/h)	240
Colors	White frame, red
Production 1989	751 (851) 18 (Indiana) 500 (906 Paso) 1,365 (750 Sport)

Raymond Roche provided Ducati its first World Superbike Championship.

the 900 Supersport replaced the 750 Sport. Both the 750 and 906 Paso were discontinued, in preparation for a sport touring replacement for 1991.

851 SP2, 851 Strada

Initially produced for the 1989 Italian Sport Production series that pitted production 750cc fours against twins of up to 1,000cc, the first 851 Sport Production was virtually indistinguishable from the Strada, but a more serious homologation model appeared for 1990, the 851 SP2.

Although still titled an 851, the 851 SP2 displaced 888cc. Apart from the increase in capacity, larger valves, and a 45mm Termignoni exhaust, the engine specifications were similar to the earlier 851 Superbike Kit. Chassis updates included an upside-down Öhlins front fork, an Öhlins shock absorber, and fully floating Brembo cast-iron front disc brakes. Although it was an expensive limited production model, the SP2 provided unparalleled handling and performance in 1990. The 851 Strada gained a dual seat with stronger rear subframe this year.

BELOW LEFT: The 888cc SP2 was one of the most impressive motorcycles available in 1990.

BELOW: The 851 Strada gained a dual seat for 1990 but was otherwise unchanged.

851 SP2 1990	Differing from the 851
Bore (mm)	94
Displacement (cc)	888
Compression ratio	10.7:1 (SP2)
Power	109 horsepower at 10,500 rpm (rear wheel)
Front suspension	42mm Öhlins
Rear suspension	Öhlins rising rate swingarm
Dry weight (kg)	188 192 (851 Strada)
Top speed (km/h)	250
Colors	White frame, red
Production	380 (SP2) 1,366 (851 Strada)

900 SUPERSPORT 1990
Differing from the 750 Sport

Power	83 horsepower at 8,400 rpm
Gears	Six-speed
Front brake (mm)	300 dual-disc
Wheels	3.50x17, 5.50x17
Tires	120/70x17 or 130/60x17, 180/55x17 or 170/60x17
Dry weight (kg)	185
Top speed (km/h)	220
Colors	Red frame, red and white
Engine numbers	From ZDM904A2C 000001
Frame numbers	From ZDM900SC 000001
Production	2,001 (900SS) 641 (400 SS Junior Japan) 12 (650 Indiana)

With its 17-inch wheels, the 900 Supersport was an improvement on the 750 Sport but still had the troublesome Weber carburetor.

900 Supersport

During 1989, the large crankcase six-speed 906 engine was adapted for the 900 Supersport. Like the 750 Sport, this was intentionally titled a 900 Supersport to create a link with the classic bevel-drive Super Sport, the theme still a large-capacity air-cooled V-twin engine in a lightweight sporting package. But the execution was flawed, and while the 17-inch wheels and larger brakes addressed some of the 750 Sport's problems, the Achilles heel remained the Weber carburetor. The exhaust system included the canister-style aluminum mufflers of the 1989 851 Strada. While the basic Marzocchi suspension was shared with the 750 Sport, the 17-inch Brembo wheels were from the 851.

This short-lived model represented an improvement over the 750 Sport, but it was still flawed, and the real successor to the bevel-drive Super Sport would come a year later. Alongside the 900 Supersport this year was a specific Japanese market 400 Supersport Junior, differing from the 900 with 16-inch Marvic composite wheels and Mikuni CV carburetors.

Doug Polen completely dominated the 1991 World Superbike Championship on the Ferracci 888.

1991

By 1991, Cagiva's overhaul of Ducati's range finally came to fruition with an improved Supersport and 907ie, production increasing to 8,061. In the World Superbike Championship, the 888 Corsa was further developed and even more dominant, the 143-kilogram 888 now producing 133 horsepower at 11,500 rpm. Doug Polen and Stéphane Mertens also received factory 888s, but no one was quite prepared for Polen's domination. Polen won seventeen races to Roche's four and Mertens's two, easily winning the championship. Fifty examples of the customer 851 Racing were also produced for 1991, Davide Tardozzi riding one to victory in the European Superbike Championship.

851 SP3, 851 Strada

The 851 SP3 was similar to the SP2, identified by louder and more upswept Termignoni exhaust pipes, while higher compression pistons and a forced air intake contributed to a slight power increase. The Brembo wheels were painted black, the brake and clutch master cylinders included remotely mounted fluid reservoirs, and as with the SP2, each SP3 received a numbered plaque. A small number of higher performance 851 SPS models was also produced, these basically with a custom 851 Racing engine. New for the 851 Strada was a Showa upside-down front fork and Öhlins rear shock.

LEFT: Similar to the SP2, setting the SP3 apart were higher mufflers and black wheels.

BELOW LEFT: The 851 Strada gained a Showa upside-down fork for 1991.

851 SP3, 851 STRADA 1991
Differing from 1990

Compression ratio	11:1 (SP3) 10.5:1 (851S)
Power	111 horsepower at 10,500 rpm (SP3) 91 horsepower at 9,000 rpm (851S)
Front suspension	41mm Showa (851S)
Rear suspension	Öhlins (851S)
Dry weight (kg)	199 (851S)
Production	534 (SP3) 16 (851 SPS) 1,200 (851S)

900, 750, 400 Supersport

While many of Cagiva's early efforts after it bought Ducati were disappointing, the company made amends with the carbureted 1991 Supersport. This was a brilliant model and has justifiably become one of the classic modern Ducatis. Although derived from the 1990 version, the 1991 Supersport looked fresh and was functionally superior to its predecessor in nearly every respect. It was also a sales success and spawned a complete new family.

While the 904cc engine was similar to the 1990 version, twin constant vacuum Mikuni carburetors replaced the troublesome Weber carburetor, and while not perfect, they were an improvement only hampered by the very long inlet tracts. The frame was new, with a shorter aluminum swingarm, as was the inverted Showa front fork and Showa rear shock absorber. The new bodywork was not only more attractive, but servicing was improved as the 17.5-liter gas tank was pivoted to allow access to the battery and airbox. There were two versions, one with a full and the other with a half fairing. While the Brembo wheels were carried over from before, the larger front brakes were now standardized with the 851. The riding position was completely revised, the higher clip-on handlebars and lower footpegs providing a unique sport touring riding position and enabling the new Supersport to be an extremely competent all-round street motorcycle.

Shortly afterward, a 750 Super Sport joined the 900. This was essentially the earlier five-speed 750 Sport engine, with Mikuni constant vacuum carburetors, installed in a similar chassis to the 900. As a budget model, the 750SS received a nonadjustable front fork, a single-front disc brake, and narrower rear wheel. Sharing this basic chassis was the Japanese market 400 Supersport Junior, this powered by the same six-speed engine as for 1990.

907 I.E.

For 1991, the 907 I.E. replaced the 906 Paso. Incorporating some 851 chassis components, the 907 I.E. was now a homogeneous sport touring motorcycle that could compete with anything on offer from Germany or Japan. Other than stronger crankcases, the liquid-cooled engine was the same as the 906, now with the 851's Weber electronic fuel injection

900, 750, 400 SUPERSPORT 1991
Differing from the 1990 900SS, 750 Sport and 400SS

Carburetion	Mikuni BDST 38
Compression ratio	9:1 (750)
Power	73 horsepower at 7,000 rpm (rear-wheel 900) 60 horsepower at 8,500 rpm (750) 42 horsepower at 10,000 rpm (400)
Front suspension	41mm Showa
Rear suspension	Showa cantilever swingarm
Front brake (mm)	320 dual-disc (900) 320 single-disc (750, 400)
Rear brake (mm)	245 disc
Tires	120/70x17, 170/60x17 (900) 160/60x17 rear (750, 400)
Wheelbase (mm)	1,410
Dry weight (kg)	183 (900) 173 (750, 400)
Top speed (km/h)	220 (900) 210 (750) 180 (400)
Colors	White frame, red
Engine numbers	Continuing ZDM904A2C 000001 series (900) From ZDM750A2C 000001 (750)
Frame numbers	From ZDM900SC2 000001 (900) From ZDM750SC 000001 (750)
Production	4,308 (900) 1,872 (750) 541 (400)

OPPOSITE: The 900 Supersport was significantly improved for 1991. Full- and half-fairing types were available. This is the half-fairing version.

ABOVE: Another improved model for 1991 was the 907 I.E.

907 I.E. 1991	Differing from the 906 Paso
Carburetion	Weber I.A.W P7
Power	90 horsepower at 8,500 rpm
Front brake (mm)	300 dual-disc
Rear brake (mm)	245 disc
Tires	120/70x17, 170/60x17
Wheelbase (mm)	1,490
Dry weight (kg)	215
Top speed (km/h)	230
Colors	Red
Engine numbers	From ZDM904W2 000001
Frame numbers	From ZDM906PI2 000001
Production	1,303

system. Also shared with the 906 was the rectangular-section steel double-cradle frame, but with the 851's longer aluminum swingarm. The wheels were now 17-inch Brembo, and while the style was similar to the 906, the solid 907 I.E. screen incorporated a NACA duct, supposedly to smooth the airflow.

1992

The combination of World Superbike success and a vastly improved production lineup saw motorcycle manufacturing increase dramatically, with 12,049 built at Borgo Panigale during 1992. The 851 received a subtle restyle and racing success continued this year. Doug Polen competed in both the World Superbike and AMA Championships. The 140-kilogram factory 888 now produced 135 horsepower at 11,200 rpm, Polen winning nine races to take his second World Superbike title. Other victories on the factory 888 went to Raymond Roche and

BELOW LEFT: Doug Polen again won the World Superbike Championship in 1992.

BELOW: By taking full advantage of the regulations, the 1992 factory racer represented the quintessence of the 888 series.

The 888 SP4 carried a larger No. 1 on the fairing.

888 SPS, SP4, 851 STRADA 1992
Differing from 1991

Compression ratio	11.2:1 (SP4) 11.7:1 (SPS)
Power	120 horsepower at 10,500 rpm (SPS) 95 horsepower at 9,000 rpm (851S)
Rear suspension	Showa (851S)
Dry weight (kg)	185 (SPS) 202 (851S)
Top speed (km/h)	255 (SP4) 260 (SPS)
Production	500 (SP4) 101 (SPS) 1,896 (851S)

The 851 Strada was restyled by Pierre Terblanche for 1992. *Ducati*

Giancarlo Falappa. Again a cataloged 888 Racing was available for 1992. Only thirty-five were built, and for the first time, the 888 Racing was genuinely competitive, Carl Fogarty riding one to a World Superbike victory at Donington.

888 SPS, SP4, 851 Strada

Two 888 Sport Production models were available for 1992, the SP4 and SPS (Sport Production Special). Sharing the Pierre Terblanche—styled new bodywork with the 1992 851 Strada, the SP4 shared engine specifications of the SP3, while the SPS was virtually a Corsa with lights. The 851 Strada was also updated, with Sport Production cylinder heads and a new frame.

900 Superlight, 900, 750, 400, 350 Supersport, 900 I.E.

Consolidating on the immediate success of the 1991 900 Supersport was a limited-edition version, the 900 Superlight for 1992. While the engine and chassis were shared with the

One of the most desirable Ducatis of the 1990s was the 1992 900 Superlight with Marvic/Akront wheels. US examples were yellow.

900 SUPERLIGHT, 900, 750, 400SS, 350, 907 I.E. 1992 Differing from 1991

Front brake (mm)	320 dual-disc (907)
Dry weight (kg)	176 (SL)
Colors	White frame, red or yellow (SL) White frame, red or black (SS) Black (907)
Production	952 (900SL) 3,302 (900SS) 2,973 (750SS) 1,051 (350/400SS) 500 (907)

900 Supersport, a vented clutch cover, fully floating cast-iron front disc rotors, and larger diameter mufflers (from the 888 SP4) set the Superlight apart. Each Superlight came with a carbon-fiber front fender, solo seat, numbered plaque on the top triple clamp, and lightweight 17-inch Marvic/Akront composite wheels. All the US examples were yellow.

The 900SS and 750SS continued largely unchanged but were now offered in black as well as red. Red versions had black wheels. A 350 Sport, for the Italian market, joined the 400SS, while the 907 I.E. continued with only minor updates. The front brakes were upgraded with larger discs and black also was available. Although it offered improved power and reliability, plus more sophisticated rear suspension than the Supersport, the 907 I.E. wasn't as popular as anticipated and was discontinued during 1992. There wouldn't be a replacement sport touring Ducati until 1997.

1993

For 1993, the 851 Strada grew to 888cc but was now toward the end of its developmental life. The 888 was also now outclassed in the World Superbike Championship, and Ducati looked toward its replacement for 1994. In the meantime, production grew to 14,494, and 1993 saw the introduction of the spectacular Supermono. After an initial release at the end of 1992, it was intended to release the new 900 Monster for 1993, but production delays resulted in the first Monsters rolling off the production line in May 1993, ostensibly as 1994 models.

Carl Fogarty's performances during 1992 earned him a factory ride alongside Falappa for the World Superbike Championship, the engine growing to 926cc and producing around 144 horsepower at 11,500 rpm. This year the 926

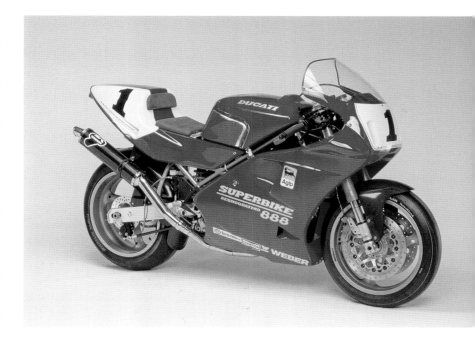

The 1993 888 Corsa was a much more highly developed machine than previous versions and similar to the factory 888 Superbike racers of the previous year. *Ducati*

One of the most beautiful Ducatis of all was the Supermono. Every component was superbly crafted.

BELOW RIGHT: The Supermono engine was basically the horizontal cylinder of the 888 Corsa, and the center of gravity was extremely low.

featured the new style bodywork, and although Fogarty won eleven races, inconsistency lost him the title. Doug Polen decided to contest the AMA Superbike National Championship, winning six of the ten rounds on the Fast by Ferracci 888 and providing Ducati its first AMA Superbike Championship. Forty-six 888 Racing machines were also built for 1993.

Supermono

One of the most interesting machines produced by Ducati was the Supermono racer of 1993–1995. This took the horizontal cylinder of an 888 Corsa, with a counterbalancing system consisting of the second con rod attached to a lever pivoting on a pin fixed in the crankcase. Titled the *doppia bielletta* (double con rod), it provided the perfect primary balance of a 90-degree V-twin. Design differences to the 888 included plain main bearings, a dry alternator, and the water pump driven off the exhaust camshaft.

SUPERMONO 1993

Engine	Single-cylinder four-stroke, liquid-cooled
Bore (mm)	100
Stroke (mm)	70
Displacement (cc)	550
Compression ratio	11.8:1
Valve actuation	Belt-driven desmodromic double overhead camshafts, four valves per cylinder
Injection	Weber I.A.W. CPU P8
Power	78 horsepower at 10,500 rpm
Gears	Six-speed
Front suspension	42mm Öhlins
Rear suspension	Öhlins cantilever swingarm
Front brake (mm)	280 dual-disc
Rear brake (mm)	190 disc
Wheels	3.50x17 and 5.00x17
Tires	310/480R17440, 155/60R17622
Wheelbase (mm)	1,360
Length (mm)	1,960
Dry weight (kg)	122
Colors	Bronze frame, red
Production	32

Housing this remarkable engine was a tubular steel frame, with an aluminum swingarm. Racing equipment extended to Marchesini three-spoke magnesium wheels, Brembo fully floating iron front discs, and racing calipers. To minimize weight, every body part was of carbon fiber, the carbon-fiber seat also acting as the rear subframe. The spectacular styling was by Pierre Terblanche, and the Supermono remains his finest effort.

888 SP5, SPO, Strada

As the official factory racers and the Sport Production series were already displacing 888cc, it was inevitable that the production Superbikes would follow suit. This occurred for 1993, the series also becoming a generic 888 rather than 851. The 888 also featured Pierre Terblanche's 1992 851 styling facelift.

The final 888 Sport Production was the 888 SP5. With the higher performance SPS engine, this continued to set the performance standard for twin-cylinder motorcycles. The frame and wheels were painted bronze, and a Showa front fork replaced the expensive Öhlins. Although there have been annual limited higher performance series since the 888, the 888 SPs were

While the 1993 SP5 lost the Öhlins front fork, the engine was upgraded to that of the 888 SPS.

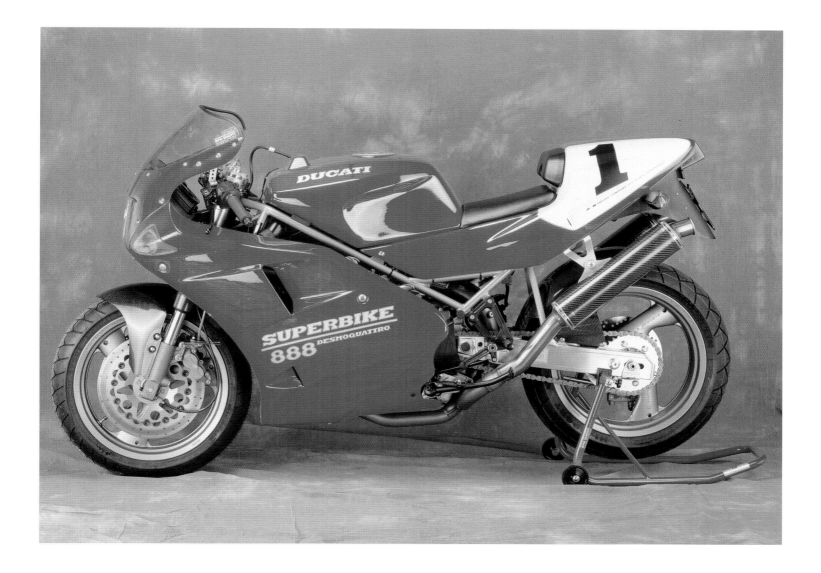

THE CAGIVA ERA | 169

The 1993 888 looked similar to the 1992 851, with new styling by Pierre Terblanche.

888 SP5, STRADA, SPO 1993
Differing from 1992

Bore (mm)	94
Stroke (mm)	64
Displacement (cc)	888
Compression ratio	11:1
Injection	Weber I.A.W. CPU P8
Power	100 horsepower at 9,000 rpm (888)
	118 horsepower at 10,500 rpm (SP5)
Front suspension	41mm Showa
Rear suspension	Showa GD
	Öhlins (SP5, SPO)
Dry weight (kg)	202 (188, SPO)
Top speed (km/h)	250 (260, SPO)
Colors	Bronze frame, red
Production	500 (SP5)
	1,280 (888)
	290 (SPO)

different. These machines were loud, hard-edged race replicas offering considerably more performance than an 851 or 888 Strada. They were also built in fewer numbers than later SPs, but they are finally achieving the recognition they deserve.

The European specification 888 Strada was very similar to that of the 1992 851 Strada, and except for an 888cc engine, most of the other specifications were shared. For the US market, the 888 was sold as the 888 SPO (Sport Production Omologato), an amalgam of the limited-production high-performance SP5 and the European-specification 888 Strada. As the SP5 was unable to pass US DOT requirements for registration, the SPO was created to homologate the 888 for AMA Superbike competition. While they were titled a Sport Production, they were more closely related to the 888 Strada than the SP5. The SPO engine was identical to that of the 888 Strada, but some chassis components were shared with the

The US 888 was the SPO, with a solo seat, upswept mufflers, and an 888 Strada engine.

SP5, including a solo seat, upswept exhaust pipes, and an Öhlins shock absorber with eccentric ride height adjustment.

900 Superlight, 900, 750, 400, 350 Supersport

As the Supersport continued to sell well, cosmetic changes predominated for 1993. Along with new graphics with silver decals, the frame was bronze. The 900 Superlight specification was downgraded for 1993, the wheels now regular Brembo.

900 SUPERLIGHT, 900, 750, 400SS, 350SS 1993
Differing from 1992

Dry weight (kg)	182 (900SL)
Colors	Bronze frame, red or yellow
Production	755 (900SL)
	3,853 (900SS)
	2,352 (750SS)
	950 (350/400SS)

Other than new graphics, the 900 Supersport was largely unchanged for 1993. *Ducati*

THE CAGIVA ERA | **171**

CHAPTER 7
GROUNDBREAKING NEW MODELS
THE 916 AND MONSTER, 1994–2002

Troy Corser won the 1994 AMA Superbike Championship on the earlier 888-based machine.

After the 1972 Imola victory, the introduction of the 916 and Monster marked the next turning point for Ducati. The Imola victory pronounced Ducati a serious mainstream manufacturer, albeit on a small scale, but the 916 and Monster were individually just as significant. As a landmark model, the 916 influenced motorcycle design for the next decade and beyond, while the Monster established a new niche that would spawn a cult of customization. While racetrack success continued with the 916, the only negative during this period was the financial difficulties Cagiva faced at this time, prompting a sale to the Texas Pacific Group in 1996.

1994

Although Ducati has produced many great motorcycles, one of the greatest was the 916. Setting a new performance and aesthetic standard, the 916 was the result of six years intensive work by Massimo Tamburini and his team at the Cagiva Research Center in San Marino. The 916 also broke from Ducati's evolutionary tradition, and apart from the engine, it was a new design. The other brilliant design introduced for the 1994 model year was Miguel Galluzzi's Monster, a naked and minimalist street bike unlike anything else available but with all the sporting credentials expected of a Ducati. The Supersport range expanded to include a 600, and production increased to 18,170 this year.

Ducati prepared the new 916 for the Superbike World Championship, retaining Carl Fogarty and Giancarlo Falappa as riders. Continually developed during the season, the 145-kilogram factory racer displaced 955cc and eventually produced 150 horsepower at 11,000 rpm. With ten victories, Fogarty narrowly won the championship. Thirty-two examples of the 1994 customer 888 Racing were also built, these similar to the 1993 926cc factory bikes. Despite the older chassis, they were still competitive, Troy Corser winning the AMA Superbike Championship on the 955cc Ferracci example. Ten Supermono racers were also built this year.

916 SP, Strada

Apart from a longer stroke, the 916 engine was similar to the 888, but around the Desmoquattro engine, Tamburini created a structure that included a stronger frame with an additional lower engine support, a sealed airbox, and a single-sided cast-aluminum swingarm. A special headlight support contained unique twin poly-ellipsoidal headlights, and the bike had specifically styled twin taillights. The exhaust system exited under the seat and was designed

OPPOSITE: Carl Fogarty rode the new 916 to victory in the 1994 World Superbike Championship. This was the last year carbon brakes were allowed in this series.

RIGHT: The 916 style was so distinctive that it has become a motorcycling icon. All 1994 916s were Monoposto.

BELOW RIGHT: Although tightly packed, the 916 was designed with ease of maintenance in mind.

916 STRADA, SP 1994
Differing from the 888

Stroke (mm)	66
Displacement (cc)	916
Compression ratio	11:1
Power	114 horsepower at 9,000 rpm 131 horsepower at 10,500 rpm (SP)
Front suspension	43mm Showa
Rear suspension	Showa rising rate mono swingarm Öhlins (SP)
Front brake (mm)	320 dual-disc
Rear brake (mm)	220 disc
Tires	120/70x17, 180/55x17 or 190/50x17
Wheelbase (mm)	1,410
Dry weight (kg)	198 192 (SP)
Top speed (km/h)	260 270 (SP)
Production	2,866 (916 Strada) 310 (916SP) 1,300 (888)

to aid aerodynamics more than provide ultimate power. The design was extremely compact and immediately became a benchmark for motorcycle style and function.

Production was initially slow due to a fire at the factory at Bologna, and most 1994 916s were assembled at Varese. Following the success of earlier Sport Production series, the 916 was also offered as an SP and S for 1994. The SP engine had larger valves, more radical camshaft timing, and a larger diameter exhaust system. The 888 Strada and SPO also continued for 1994, now with bronze wheels.

The 916SP looked similar to the 916 Strada but offered considerably more performance.

900, 600 Monster

As with some earlier efforts, the 900 Monster was a synthesis of components of various existing models but concocted into a considerably more effective package than most previous endeavors. While the engine was basically the 900 Supersport, the more sophisticated chassis with rising rate rear suspension was derived from the 851/888. The instrumentation and bodywork were new, the minimalist instrument panel initially without a tachometer.

A lightweight 600 Monster also became available during 1994, sharing the five-speed engine from the newly introduced 600 Supersport in a chassis similar to the 900 Monster. The 600 Monster was a popular model in Europe, where it was offered with a lower output engine for specific insurance categories, as was a specific 400 version for Japan.

900 Superlight, SP, CR, 900, 750, 600, 400 Supersport

Updates to the 1994 Supersport lineup included frame-matching bronze wheels, a 900CR for the United States, and smaller 600. The 900 Supersport and Superlight received an improved front fork and stronger swingarm, and after the 750SS's dismal failure in the United States, a new model mating the 900 Supersport engine with the more basic 750 Supersport chassis was offered. The 900CR and 750SS received dual front disc brakes.

BELOW LEFT: With the 900 Monster, Ducati successfully created a new niche in the market. The Monster was so successful the same basic style continued for more than a decade.

BELOW: Joining the 900 Monster during 1994 was the 600. This had a single front disc brake and basic suspension. *Ducati*

900, 600 MONSTER 1993–1994
Differing from the 900SS and 600SS

Rear suspension	Boge rising rate swingarm
Rear tire	170/60x17
Wheelbase (mm)	1,430
Dry weight (kg)	184 (900) 174 (600)
Top speed (km/h)	190 (900) 175 (600)
Production (1993–1994)	7,262 (900) 4,195 (600/400)

The 600 Supersport was also new for 1994. *Ducati*

After 1992, there were no more Superlights for the United States, but 1994 saw the introduction of the 900SS SP (Sport Production). The 900SS SP was an amalgam of the Supersport and the Superlight, and it was essentially a fully faired Superlight with a dual seat and lower exhaust pipes. Like the Superlight, they carried numbered plaques on the top triple clamp and included a carbon-fiber clutch cover and front and rear fenders, fully floating rear disc brake, and floating cast-iron front disc rotors.

New for 1994 was the 600 Supersport. While the crankcases, five-speed gearbox, and cylinders were similar to the earlier 600SL Pantah, the 600 Supersport included the small-valve 400 cylinder heads. There were two versions (33 and 53 horsepower), both with a two-into-one exhaust, sharing the single front disc brake 400 chassis.

900, 750, 600, 400 SUPERSPORT, 900SL, SP, CR 1994 Differing from 1993

Bore (mm)	80 (600)
Stroke (mm)	58 (600)
Displacement (cc)	583 (600)
Compression ratio	10.7:1 (600)
Power	53 horsepower at 8,250 rpm (600) 33 horsepower at 7,500 rpm (600)
Front brakes	320 dual-disc (900CR, 750SS)
Dry weight (kg)	186 (900) 176 (750) 172 (600)
Top speed (km/h)	220 (900) 200 (750) 190 (600)
Production	2,816 (900SS, CR) 1,471 (900SL, SP) 1,008 (750) 2,500 (600/400)

1995

With Cagiva now canvassing a buyer for Ducati, this year was one of financial instability. As a result, development was limited to the 748 Desmoquattro, 916 Senna, and Biposto 916; other models continued largely unchanged. But despite the financial problems, production rose to 20,989. This year saw continued success in World Superbike, the 155-kilogram 955 now producing 154 horsepower at 12,000 rpm. The factory racer was particularly effective this year, providing an

Fogarty completely dominated the 1995 World Superbike Championship on the factory 955.

optimum balance that enabled Fogarty to win thirteen races, sealing the championship with two rounds to spare.

The 1995 customer 916 Racing was closely patterned on Fogarty's 1994 factory bike. Sixty 916 Racing machines were available for teams throughout the world, and they were very competitive, Steve Hislop winning the British Superbike Championship. A second series Supermono was also produced this year, now including a slightly larger engine and revised suspension. A production Supermono was always promised but never happened.

SUPERMONO 1995
Differing from 1994

Bore (mm)	102
Displacement (cc)	572
Power	81 horsepower at 10,000 rpm
Production	25

916 Senna, 916/748SP, Strada, Biposto

Updates for 1995 saw the 916 Strada become a Biposto (dual seat) in all markets except the United States and the introduction of a new Weber electronic injection system. Joining the 916 lineup was the limited-edition Senna, created in memory of Brazilian Formula One driver Ayrton Senna. Only a few months before his death in 1994, Senna agreed to the production of a limited-edition 916 Senna, with the profits going to the Senna Foundation. The 1995 Senna was ostensibly a standard monoposto 916 in gray, with red wheels.

As the Desmoquattro started its life as a 748, and the cylinder heads were designed around a smaller bore, it wasn't surprising to see the 748 introduced for 1995. There were initially three versions, the Strada, Biposto, and Sport Production.

Based on the larger crankcase six-speed engine, the dimensions were the same as the earlier 748. The combination of less internal reciprocating engine weight, a lower profile front, and narrower rear tire, in the opinion of many, provided superior steering and handling to the 916. Also available this year was the striking yellow 748 Sport Production. It was a homologation machine intended for the expanding Supersport racing category in Europe, where the 748cc twin could compete against 600cc four-cylinder machines.

GROUNDBREAKING NEW MODELS | 177

TOP: The 916SP was little changed for 1995.

BELOW: The striking 748SP was also new for 1995.

916, 748 STRADA, BIPOSTO, SP, 916 SENNA 1995 Differing from 1994

Bore (mm)	88 (748)
Stroke (mm)	61.5 (748)
Displacement (cc)	748 (748)
Compression ratio	11.5:1 (748), 11.6:1 (748SP)
Injection	Marelli 1.6M
Power	98 horsepower at 11,000 rpm (748) 104 horsepower at 11,000 rpm (748SP)
Rear suspension	Öhlins (Senna, 748SP)
Tires	120/60x17, 180/55x17 (748)
Dry weight (kg)	200 (748 Strada) 202 (748 Biposto) 198 (748SP)
Top speed (km/h)	240 (748) 250 (748SP)
Production	401 (916SP) 2,543 (916S/Bip) 301 (Senna) 2,562 (748)

The 1995 Senna I was virtually a standard 916 with some SP equipment.

900, 600 Monster, 900 Superlight, SP, CR, 900, 750, 600, 400 Supersport

Only cosmetic updates distinguished the 1995 Monster and Supersport from earlier models, the frame now 916-gold, with matching wheels.

1996

During 1996, it was obvious that there were difficulties at Borgo Panigale. Motorcycle production dropped to only to 12,509, with partially built motorcycles lining the factory floors as suppliers weren't being paid. Amongst many rumors of a buyout, eventually a new company, Ducati Motor, was created as a joint venture between the US-based Texas Pacific Group and Cagiva. Now independent, Ducati also had to set up its own commercial and research and development departments, and new models were no longer designed at the Cagiva Research Center in San Marino.

Production difficulties aside, the World Superbike racing program continued to be well supported and given high priority. With the minimum weight increased to 162 kilograms, the factory bikes now displaced 996cc and produced 157 horsepower at 11,800 rpm. John Kocinski rode for Virginio Ferrari's Ducati Corse, and Troy Corser for Promotor. Corser won seven races, taking the championship.

900, 600 MONSTER, 900, 750, 600, 400 SUPERSPORT, 900SL, SP, CR 1995
Differing from 1994

Front suspension	40mm Marzocchi (900CR, 750)
Rear suspension	Sachs-Boge (900CR)
Production	3,657 (M900)
	3,670 (M600/400)
	3,569 (900SS, CR)
	1,330 (900SL, SP)
	1,550 (750)
	737 (600/400)

GROUNDBREAKING NEW MODELS | 179

Troy Corser won the 1996 World Superbike Championship on the Promotor 996.

RIGHT: The 1996 factory Superbike unclothed.

916SPA, 916SP3, 916/748 STRADA, BIPOSTO, 748SP 1996 Differing from 1995

Bore (mm)	96 (SPA)
Displacement (cc)	955 (SPA)
Production	54 (SPA)
	497 (SP3)
	1,844 (916S/Bip)
	1,560 (748)

955SP, 916SP3, 916 Strada, Biposto, 748SP, Strada, Biposto

The 916SP became the SP3 this year, ostensibly identical to the 1995 916SP except with fewer carbon-fiber components. There was no Senna this year, but to homologate the 955cc engine for AMA racing, the 955cc 916SPA was available in the United States. The 916 and 748SP and Biposto were virtually identical to 1995.

900, 750 600 Monster, 900 Superlight, SP, CR, 900, 750, 600, 400 Supersport

The 750 Monster was the only new model for 1996, ostensibly the single-disc 600 Monster chassis with a five-speed 750 Supersport engine. Nine hundred engines received new crankcases, but all other two-valve models continued unchanged.

ABOVE LEFT: The 916 Biposto continued largely unchanged for 1996. *Ducati*

BELOW LEFT: The only new model for 1996 was the 750 Monster, ostensibly a 750cc engine in the 600 Monster chassis. *Ducati*

900, 750 600 MONSTER, 900, 750, 600, 400 SUPERSPORT, 900SL, SP, CR 1996
Differing from 1995 and the 750SS

Front suspension	40mm Marzocchi (M900)
Dry weight (kg)	178 (M750)
Top speed (km/h)	190 (M750)
Production	2,186 (M900)
	1,306 (M750)
	1,762 (M600/400)
	1,328 (900SS, CR)
	981 (900SL, SP)
	60 (750 SS)
	663 (600/400SS)

1997

Federico Minoli took over as managing director of Ducati Motor in April, and this year saw the introduction of the 996cc 916SPS and the first of the new Sport Touring family, the ST2. With the financial security provided by the new ownership, production soared to 27,051. After a one-year hiatus, Carl Fogarty returned to Virginio Ferrari's factory Superbike team, but 1997 was an indifferent year. Although the power was now 168 horsepower at 11,300 rpm and reliability was improved, the overall balance was inferior. Although he won six races, Fogarty lost the title.

916SPS, 916 Senna, Strada, Biposto, 748SP, Strada, Biposto

The most significant new model for 1997 was the 916SPS. Although it still carried the generic 916 title, the 916SPS displaced 996cc, with new crankcases for World Superbike homologation. Other updates included larger inlet valves and intake ports, and 50mm exhaust header pipes. Another Senna series, the Senna II, was also built for 1997, these virtually identical to the earlier version but with lighter gray colors. The existing 916 and 748 models continued unchanged.

For 1997, the 916SP grew to 996cc and became the 916SPS. Visually similar to the 916SP, it provided considerably more performance.

Other than new colors, the 1997 Senna II was virtually identical to the 1995 Senna I.

ST2

Continuing where the 906 and 907 I.E. left off, the ST2 was a more conventional Sport Touring model. The 944cc water-cooled two-valve engine was an evolution of the 907, while the frame was based on the 900 Monster with a 916-style rear suspension linkage. Styled by Miguel Galluzzi, the level of equipment was also higher than earlier sport touring Ducatis, with side panniers a factory option.

916SPS, SENNA, 916/748 STRADA, BIPOSTO, 748SP 1997 Differing from 1996

Bore (mm)	98 (916SPS)
Displacement (cc)	996 (916SPS)
Compression ratio	11.5:1 (916SPS)
Power	134 horsepower at 10,500 rpm (916SPS)
Dry weight (kg)	190 (916SPS)
Production	404 (916SPS) 301 (Senna) 3,814 (916S/Bip) 3,243 (748)

The ST2 initiated a new family of sport touring Ducatis.

ST2 1997
Differing from the 907 I.E.

Bore (mm)	94
Displacement (cc)	944
Compression ratio	10.2:1
Injection	Marelli 1.6M
Power	83 horsepower at 8,500 rpm
Front suspension	43mm Showa
Rear suspension	Showa rising rate swingarm
Wheelbase (mm)	1,430
Dry weight (kg)	209
Top speed (km/h)	215
Production	4,324

GROUNDBREAKING NEW MODELS

The final year for the older-style classic Supersport was 1997. Now with new graphics, a 900 is in the foreground with a 750 behind.

900, 750, 600, 400 MONSTER, 900SP/CR, 900, 750, 600, 400 SUPERSPORT 1997
Differing from 1996

Power	66.6 horsepower at 7,000 rpm (M900)
Production	3,357 (M900)
	2,352 (M750)
	4,258 (M600/400)
	1,411 (900SS, CR)
	300 (900SP)
	1,146 (750 SS)
	1,001 (600/400SS)

900, 750, 600, 400 Monster, 900 SP/CR, 900, 750, 600, 400 Supersport

In an endeavor to improve low-rpm torque, the 900 Monster engine was detuned, with smaller valves and a milder camshaft. Also new this year was a standard handlebar fairing, but the other Monsters were unchanged. While waiting for a replacement Supersport, the existing 900 Supersport continued with new graphics and minor updates. As the 900 Superlight was discontinued, the rear brake was now the Superlight fully floating arrangement. While still a competent motorcycle, by 1997 the Supersport was deemed antiquated and obsolete, and it was time for a change. The US 900SS SP continued this year, as did the 900CR and 750SS, all with new graphics.

1998

During 1998, TPG bought the remaining 49 percent of Ducati from Cagiva. TPG's buyout immediately became evident with Massimo Vignelli's new logo, an increase in the workforce to 714 employees, and the establishment of the Ducati Performance subsidiary. New models included the 900 Supersport, and overall production increased slightly to 27,970. A concerted effort was made this year to win the World Superbike Championship, with prior champions Fogarty and Corser riding factory machines for separate teams. The factory 996 was significantly updated, with a new frame and engine management system, and produced around 163 horsepower at 12,000 rpm, enough for Fogarty to win his third World Superbike Championship.

Carl Fogarty won his third World Superbike Championship in 1998 on the factory 996.

GROUNDBREAKING NEW MODELS

ABOVE: The 916SPS Fogarty was produced in limited numbers during 1998 to homologate a new racing frame.

RIGHT: Models in 1998 had no gas tank decals. This is a US-spec 916 Monoposto.

916SPS, 916 Senna, Strada, Biposto, 748SPS, Strada, Biposto

Apart from new graphics, the existing 916 and 748s continued much as before. To make it more suitable for World Supersport racing, the 748SPS replaced the 748SP, and a special 916SPS Fogarty Replica was built to homologate the new World Superbike frame. The first series Fogarty Replicas were primarily for the UK market. This year also saw the final series of 916 Sennas, and a limited number of 748 Racing machines were made available to select teams.

916/748SPS, SENNA, 916/748 STRADA, BIPOSTO 1998 Differing from 1997

Dry weight (kg)	194 (748SPS)
Production	958 (916SPS)
	202 (916SPS Fogarty)
	300 (Senna)
	2,827 (916S/Bip)
	3,431 (748)

The final 916 Senna appeared for 1998, still much as prior versions, apart from colors and graphics.

Continuing the Superlight tradition, the 900 Supersport FE was the last of the carbureted Supersports.

900 SUPERSPORT, 900SS FINAL EDITION 1998
Differing from 1997

Injection	Marelli 1.5M (900SS)
Power	80 horsepower at 7,500 rpm (900SS)
Front suspension	43mm Showa (900SS)
Rear suspension	Showa cantilever swingarm (900SS)
Wheelbase (mm)	1,395 (900SS)
Dry weight (kg)	188 (900SS)
Top speed (km/h)	225 (900SS)
Production	4,406 (900SS) 900 (FE)

900 Supersport, 900SS Final Edition

Replacing the long-running 900 Supersport for 1998 was a new fuel-injected version. Designed by Pierre Terblanche, as it was considered a traditional model, the engine remained air-cooled, with a similar frame and cantilever swingarm as before, but the suspension was a newer generation Showa. Also available this year was a 900SS Final Edition, similar to the previous Superlight.

The 900 Supersport introduced in 1998 looked very different to its predecessor, but it continued the classic theme.

900, 750, 600 MONSTER, ST2 1998
Differing from 1997

Front suspension	41mm Showa (M900)
Front brakes	320 dual-disc (M750)
Production	3,297 (M900)
	1,804 (M750)
	5,596 (M600)
	2,494 (ST2)

900, 750, 600 Monster, ST2

The delayed production of the ST2 during 1997 saw it continue into 1998 with only a few updates (notably the decals), but the Monster lineup now included the 900S and 900 Cromo (with chrome-plated gas tank). The basic low output 900 Monster continued (without a small fairing), while the 900S included the earlier 73-horsepower engine and an adjustable Showa front fork. The 750 Monster received dual-disc front brakes, but perhaps the most significant new model was the entry-level 600 Monster Dark. Designed to expand the customer base, this would become the most popular model in Italy.

1999

This year the 996 replaced the 916, the ST4 joined the Sport Touring lineup, and a fuel-injected 750 Supersport was introduced. Motorcycle production rose to 34,657. For 1999, there was only one officially supported factory World Superbike team, Ducati Performance, with riders Carl Fogarty and Troy Corser. Although the 168-horsepower 996 Racer was little changed from the previous year, evolutionary development saw the balance return, as did its dominance. Fogarty won eleven races and Corser three, while on a customer 996 Racing Special, Troy Bayliss won the British Superbike Championship and Steve Martin the Australian Superbike Championship.

The 1998 M900S had higher specification brakes than the 900 Supersport motor.

Fogarty was totally dominant in 1999, easily winning the World Superbike Championship.

The 1999 factory 996 was extremely well developed.

Replacing the 916 for 1999 was the 996, with a larger engine and improved brakes.

996/748SPS, BIPOSTO, 996S 1999
Differing from 1998

Bore (mm)	98
Displacement (cc)	996
Compression ratio	11.5:1
Dry weight (kg)	198 (Biposto), 195 (Strada)
Production	6,929 (996)
	2,228 (748)

996SPS, 996 Strada, Biposto, 748SPS, Biposto

Replacing the 916 this year was the 996, now sharing the crankcases and pistons of the SPS. The SPS also became a 996 but was little changed, as were the two 748s.

ST4, ST2

As the 916 evolved into the 996, the 916cc engine made its way into the ST4. The ST4 was an amalgam of the Desmoquattro engine in the ST2 chassis, with the front cylinder head shortened to maintain adequate front wheel clearance. Engine specifications were identical to the 916, and the chassis was virtually identical to the ST2.

BELOW LEFT: The first ST4 looked almost identical to the ST2, but underneath the bodywork was a 916cc Desmoquattro engine.

ST4, ST2 1999 Differing from 1998

Bore (mm)	94 (ST4)
Stroke (mm)	66 (ST4)
Displacement (cc)	916 (ST4)
Compression ratio	11:1 (ST4)
Valve actuation	**Belt-driven desmodromic double overhead camshafts, four valves per cylinder (ST4)**
Power	107 horsepower at 9,000 rpm (ST4)
Dry weight (kg)	215 (ST4)
Top speed (km/h)	245 (ST4)
Production	3,476 (ST4, ST2)

SUPERSPORT AND MONSTER 1999
Differing from 1998

Bore (mm)	88 (750SS)
Stroke (mm)	61.5 (750SS)
Displacement (cc)	748 (750SS)
Compression ratio	9:1 (750SS)
Injection	Marelli 1.5M (750SS)
Power	64 horsepower at 8,250 rpm (750SS)
Wheelbase (mm)	1,405 (750SS)
Top speed (km/h)	205 (750SS)
Production	2,265 (900SS) 2,760 (750SS) 6,393 (M900) 3,037 (M750) 7,808 (M600/400)

Joining the 900 Supersport for 1999 was the smaller 750, still with a five-speed gearbox. There was a choice of full or half fairing (as on this example).

Supersport and Monster

A 750 Supersport joined the new 900 for 1999. This included the five-speed 750 Monster engine, but with Weber Marelli electronic fuel injection. The 1999 model year was also characterized by the proliferation of Monsters, now including fifteen models. The Monster 900, 900 Dark, 900S, 900 Cromo, and 900 California received the 74-horsepower engine, while the Monster 900 City and 900 City Dark included the small-valve 71-horsepower engine. With its black frame and matte black tank and fenders, the Dark represented the austere Monster, and alongside the 900 was the Monster 750 Dark, 750 City, 750 City Dark, Monster 600 City, and 600 City Dark.

2000

Ducati celebrated the new millennium with the Internet sale of Pierre Terblanche's MH900e. This was similar to his 1998 concept version, but due to production delays, most were eventually built in 2001. In its final year before replacement by the Testastretta, the 996 Desmoquattro was offered as one more limited edition, while the 748R was introduced as a World Supersport homologation model. For the first time since 1993, the Monster was revised, and while overall motorcycle production increased to 39,549, the strongest growth was now in apparel and accessories.

In the World Superbike Championship, former AMA Champion Ben Bostrom joined Carl Fogarty in the official Team Ducati Infostrada. Fogarty crashed at Phillip Island, the injuries ending his career, and Troy Bayliss eventually took his place. While the factory 996 now produced 173 horsepower at 12,000 rpm, the team was ill prepared for Fogarty's withdrawal from the championship, and Bayliss only won two races. A further ten 996 Racing Specials were produced for selected teams in World Superbike and national championships, Neil Hodgson winning the British Superbike Championship.

The last of several limited-edition series of the 996SPS was the Pista of 2000.

996SPS, S, BIPOSTO, 748R, S, ECONOMY 2000 Differing from 1999

Power	97 horsepower at 11,000 rpm (748S, Economy)
	106 horsepower at 11,000 rpm (748R)
Front suspension	43mm Öhlins (996SPS)
Rear suspension	Öhlins (996SPS, 748R)
	Sachs-Boge (748 Economy)
Dry weight (kg)	187 (996SPS)
	192 (748R)
	196 (748 Economy)
Top speed (km/h)	255 (748R)
Production	5,000 (996)
	7,791 (748)

The 2000 748R incorporated many features previously reserved for racing-only machines.

996SPS, 996S, 996 Biposto, 748R, 748S, 748 Economy

The 996SPS received an Öhlins front fork, and this year saw the final Fogarty Replica, the 996 Factory Replica 2, or Pista (circuit). A 996S was available in the United States this year, ostensibly a 996SPS with a stock 996 motor, while the highest specification production Ducati to date, the 748R, replaced the 748SPS. This included the racing shower injection system and larger capacity airbox. The 748S and basic 748 Economy completed the three-model 748 range.

Monster, ST4, ST2, 900, and 750 Supersport

Monsters were important to Ducati, comprising 43 percent of sales during 1999, and Ducati embarked on a program of model expansion, offering an almost bewildering range of Monsters in 900, 750, and 600cc for 2000. Variations included the Monster Dark, City, Metallic, and Special, and all versions received a larger inverted Showa front fork, restyled taillight, fuel tank, wheels and fenders, and upgraded front brakes. The Monster 900 was now fuel injected, while the 600 and 750 continued with Mikuni carburetors.

Apart from cosmetics, there were few changes to the Sport Touring models for 2000. The ST4 brakes were upgraded with thicker discs to differentiate the model from the ST2, and it received a larger rear tire. There were also only minimal updates to the 900 and 750 Supersport this year, primarily a higher handlebar and windshield to improve rider comfort.

2001

This year saw the release of the second-generation Desmoquattro engine, the Testastretta or "narrow head." Initially powering the limited-edition 996R, the Testastretta would eventually replace the Desmoquattro. In the meantime, the 996 Desmoquattro engine made it to the ST4S, with the 916 unit now powering the Monster S4 and Limited-Edition Fogarty S4. The MH900e also went into production, and the total number of motorcycles built in 2001 increased to 40,602.

Ducati Corse supported two teams in World Superbike with the new 998 Testastretta. Ruben Xaus joined Bayliss in Team Ducati Infostrada, with Ben Bostrom in a separate L&M-sponsored team. The Testastretta engine featured a narrower 25-degree included valve angle and a larger bore and shorter stroke; the factory racers produced a claimed 176 horsepower at 12,000 rpm.

Bayliss was immediately happier with the 998, and with six wins, he took the World Superbike Championship. Bostrom also had an excellent season, winning five races in succession. The customer 996RS was similar to the 2000 factory bike, retaining the earlier 996cc Desmoquattro engine. In Britain, John Reynolds, Sean Emmett, and Steve Hislop were unbeatable, Reynolds eventually taking the British Superbike Championship.

MONSTER, SPORT TOURING, SUPERSPORT
Differing from 1999

Carburetion	Marelli 1.5M (M900 I.E.)
Power	78 horsepower at 8,000 rpm (M900 I.E.)
Front suspension	43mm Showa (Monster)
Rear tire	180/55x17 (ST4)
Top speed (km/h)	210 (M900 I.E.)
Production	6,393 (M900) 3,910 (M750) 8,179 (M600) 2,738 (ST4, ST2) 2,338 (900SS) 1,687 (750SS)

TOP LEFT: The 900 Monster was mildly restyled by Pierre Terblanche for 2000 and received fuel injection.

LEFT: Troy Bayliss won the 2001 World Superbike Championship on the new 998 Testastretta.

The 996R was the homologation model for the Testastretta engine. Also homologated was the smooth fairing without air scoops.

996R

The Testastretta was homologated for World Superbike with the limited production 996R. Replacing the 996SPS, the 996R was built as a limited production of 500 units, and due to street homologation difficulties, all US examples came without lights and stands, although the wiring remained in place. With the 996R only a limited-edition model, the previous 123-horsepower 996SPS engine now powered the 996S, this filling the lineup between the 996R and 112-horsepower 996. A specific US 996S was also available this year, with an Öhlins front fork, Ducati Corse decals, and the lower horsepower engine. Other than the 748R's Öhlins fork, the three 748s were little changed.

Monster, Sport Touring, Supersport

The Monster range was expanded to include the long-awaited Desmoquattro Monster S4. This was not a variant of the existing two-valve versions, but an adaptation of the ST4, sharing its basic 916 engine and frame architecture with the sport touring model. In honor of the company's most successful racer Carl Fogarty, during 2001 Ducati offered a limited-edition Monster S4 Fogarty. Ostensibly a standard Monster S4, stylist Aldo Drudi provided new color-coded bodywork, and it was available with a high-level Termignoni exhaust system. While the basic two-valve Monster was unchanged for 2001, the lineup included four 900s and three 600s and 750s: the Dark, Metallic, Cromo, and the 900 Special. The 400, 600, and 750 Monster still featured carburetors and all models included a tachometer. For 2001, the 996cc ST4S joined the Sport Touring lineup alongside the ST2 and ST4, while the Supersport range expanded to include an entry-level 750 Sport.

996R, 996S, 996, 748R, 748S, 748 2001
Differing from 2000

Bore (mm)	100 (996R)
Stroke (mm)	63.5 (996R)
Displacement (cc)	998 (996R)
Compression ratio	11.4:1 (996R)
Injection	Marelli 5.9M (996R)
Power	135 horsepower at 10,200 rpm (996R)
Front suspension	43mm Öhlins (996R, 748R)
Rear suspension	Öhlins (996R, 996S, 996)
Dry weight (kg)	185 (996R)
Production	5,762 (996) 4,359 (748)

A limited-edition Monster S4 offered for sale over the Internet during 2001 was the Foggy Monster S4.

The ST4S was powered by the 996cc Desmoquattro motor and included higher quality chassis components.

MONSTER S4, FOGARTY S4 2001
Differing from the 2000 Monster

Engine	Twin-cylinder four-stroke, liquid-cooled
Bore (mm)	94
Stroke (mm)	66
Displacement (cc)	916
Compression ratio	11.0:1
Valve actuation	Belt-driven desmodromic double overhead camshafts, four valves per cylinder
Injection	Marelli 5.9M
Power	101 horsepower at 8,750 rpm 110 horsepower at 9,750 rpm (Foggy)
Tires	120/70x17, 180/55x17
Wheelbase (mm)	1,440
Dry weight (kg)	193 (189, Foggy)
Top speed (km/h)	225
Production	8,183 (S4 2000, 2001) 2,782 (M900) 1,978 (M750) 8,823 (M600/400)

ST4S 2001 Differing from the ST4 2000

Bore (mm)	98
Stroke (mm)	66
Displacement (cc)	996
Compression ratio	11.5:1
Injection	Marelli 5.9M
Power	117 horsepower at 8,750 rpm
Rear suspension	Sachs Boge (ST4) Öhlins (ST4S)
Dry weight (kg)	212
Top speed (km/h)	255
Production	3,734 (ST4S, ST4, ST2)

750 SPORT 2001
Differing from the 750SS 2000

Front brake (mm)	320 single-disc
Wheelbase (mm)	1,410
Dry weight (kg)	181
Production	1,515 (900SS) 2,974 (750SS, 750S)

Mike Hailwood 900 Evoluzione

Inspired by the 900NCR racers of the late 1970s, general production of Terblanche's limited-edition MH900e commenced during 2001. Powered by the fuel-injected 900 Supersport engine, the tubular-steel-trellis frame was specific to the MH900e, as was the single-sided swingarm. As on the 1998 prototype, the engine featured styled sump covers, and the graphics were the earlier Giugiaro type. While the specification was unexceptional, the MH900e was one of Terblanche's more spectacular styling efforts, and despite the relatively large number produced, it remains one of the more collectable modern Ducatis.

2002

During 2002, Ducati embarked on the development of a MotoGP racer, unveiling the Desmosedici prototype at the end of May. New models this year included the interim 998 Testastretta and updates to the entire Monster lineup. Motorcycle sales topped 39,607 during 2002, the Monster predominating with 19,600 sold. Total production was 39,534.

Ducati Corse continued to update the Factory Superbike, the 998F02 now an interim model until the advent of the new 999. With a second-generation Testastretta engine, with a larger bore and shorter stroke, the power was increased to 188 horsepower at 12,500 rpm. This year Bayliss narrowly failed in his quest for another World Superbike Championship, but Ducati had more success in the British Superbike Championship, Steve Hislop winning the championship on a customer 998 Testastretta.

998R, 998S, 998, 748R, 748S, 748

For 2002, the 998cc Testastretta engine powered all the large displacement Superbikes, now with the generic title 998. Only the 998R featured the short-stroke 999cc motor, while the 748 range was unchanged except for colors and decals. Heading the range was the spectacular 999cc 998R, with the larger bore and shorter stroke engine of the World Superbike 998F02 racer. Combining the timeless 916 style with the newer generation Testastretta engine, as the final example of the 916-derived limited-edition homologation models, the rare 998R was arguably the finest.

Another Terblanche effort was the distinctive MH900e. While the engine was stock 900 Supersport, most components were unique to the MH900e.

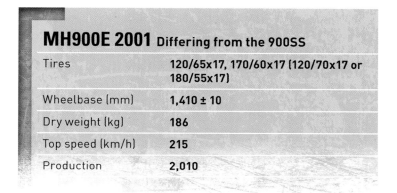

MH900E 2001	Differing from the 900SS
Tires	120/65x17, 170/60x17 (120/70x17 or 180/55x17)
Wheelbase (mm)	1,410 ± 10
Dry weight (kg)	186
Top speed (km/h)	215
Production	2,010

The style of the MH900e gas tank and fairing was intentionally designed to create an association with the NCR racers of the late 1970s.

Ben Bostrom's 998F02 World Superbike racer was an interim model with the 999cc short-stroke engine. *Ducati*

Somewhat down the scale in specification below the 998R, the 998S was ostensibly the 2001 996R engine, in the 998 chassis. Powering the base 998 was a lower horsepower Testastretta. As the higher horsepower versions weren't street legal in America, US 998Ss also received the lower horsepower 998 motor but included an Öhlins front fork. Also for 2002 Ducati released two limited-edition 998S models: the Bayliss 998S and Bostrom 998S. Both included an Öhlins front fork and colors replicating the racers, the Bostrom built in three series of 155 (his racing number).

Monster, Sport Touring, Supersport

The three-model Sport Touring range was much as before, as was the Supersport lineup, except for an additional model: the 900 Sport. This expanded the range to four models, and to differentiate it from the 900 Sport, the 900 Supersport received an Öhlins rear shock absorber and new aluminum swingarm.

BELOW RIGHT: The spectacular 998R was the final limited-edition series based on the 916.

998, R, S, BOSTROM, BAYLISS,
Differing from 2001

Bore (mm)	104 (998R)
Stroke (mm)	58.8 (998R)
Displacement (cc)	999 (998R)
Compression ratio	12.3:1 (998R)
Power	139 horsepower at 10,200 rpm (998R) 136 horsepower at 10,200 rpm (998S) 123 horsepower at 9,750 rpm (998, 998S US)
Dry weight (kg)	183 (998R) 185 (Bostrom US) 187 (998S, Bostrom, Bayliss) 198 (998)
Top speed (km/h)	280 (998R) 270 (998S)

The 998S Bostrom garishly replicated Ben Bostrom's factory Superbike racer. *Ducati*

BELOW: The 2002 620S I.E. had an updated engine, dual-front disc brakes, and a small windshield. *Ducati*

ABOVE: Another limited-edition 998S was the Bayliss, also decked out in full Superbike replica colors.

BELOW Also new for 2002 was the more basic 900 Sport. *Ducati*

By 2002, the Monster had become the most important model in Ducati's history, with sales topping 100,000 and symbolizing Ducati even more than the classic 916 to 998. The Monster range was expanded to include five different displacement versions, and apart from the base 600 Dark, the entire family was redesigned along the lines of the S4. This included new electronic fuel injection and the S4 frame. New this year were the 620 I.E. and 620S I.E. Monster, the five-speed 620 derived from the 600, joined by the 750 I.E. and 750S I.E. for the United States. As the model was about to be replaced, the M900 I.E. and M900 I.E. Dark were the only 900 Monsters in the lineup this year.

900 SPORT 2002
Differing from the 900SS 2001

Front suspension	43mm Marzocchi (900 Sport)
Rear suspension	Sachs (900 Sport) Öhlins (900SS)
Wheelbase (mm)	1,405

900, 750, 620 I.E. MONSTER 2002
Differing from 2001

Bore (mm)	80 (620)
Stroke (mm)	61.5 (620)
Displacement (cc)	618 (620)
Compression ratio	10.5:1 (620)
Injection	Marelli 5.9M
Power	60 horsepower at 9,500 rpm (620) 64 horsepower at 8,750 rpm (750)
Front suspension	43mm Marzocchi (620)
Rear suspension	Sachs-Boge (620)
Front brake (mm)	320 dual-disc (620)
Wheelbase (mm)	1,440
Dry weight (kg)	177 (620) 179 (750) 189 (900)
Top speed (km/h)	185 (620) 195 (750)

CHAPTER 8
CONFUSED DIRECTIONS
THE 999 AND SPORT CLASSIC, 2003–2006

The first version of the Desmosedici MotoGP racer. It was surprisingly competitive from the outset. *Ducati*

Neil Hodgson completely dominated the 2003 World Superbike Championship on the new 999. *Ducati*

Ducati headed in a new direction in 2003, the controversial 999 replacing the iconic 916/996/998. While the first stage in this evolutionary replacement occurred with the introduction of the Testastretta engine in the 996R and 998, it was inevitable that a new model would replace the aging 916. Although still more evolutionary than revolutionary, the 999 was considerably updated over the 998, noticeably with Pierre Terblanche's divisive styling. Although still successful in the World Superbike Championship, the 999 would ultimately fail in the market and result in Ducati introducing a replacement only three years after its introduction. This short, confused period ended with the TPG selling its shareholding to the Italian company Investindustrial early in 2006.

2003

This year not only saw the release of the 999 and Multistrada, but the 1000 and 800 Supersport and Monster replaced the previous 900 and 750. Unfortunately, resistance to the 999 saw motorcycle production decrease to 38,417. While response to the 999 was underwhelming, Ducati surprised everyone with the immediate competiveness of the Desmosedici in the highest level of motorcycle racing, MotoGP. Under the direction of Filippo Preziosi, the 90-degree V-four combined established technology (desmodromic valves and a tubular-steel frame) with state-of-the-art aerodynamics and suspension. The liquid-cooled 16-valve 86x42.6mm engine produced more than 220 horsepower at 16,000 rpm, and riding the Desmosedici during 2003 were Loris Capirossi and Troy Bayliss. When Capirossi won at Catalunya, he provided Ducati its first Grand Prix victory since 1959. Capirossi finished fourth in the championship, with Bayliss sixth, and Ducati ended second in the manufacturer's championship—a phenomenal first time effort for such a small manufacturer.

Loris Capirossi on the 2003 D16. He won one Grand Prix and finished fourth in the world championship. *Ducati*

With all the other manufacturers deciding to concentrate on MotoGP, Ducati fielded the only factory-supported team in the World Superbike Championship. With thirteen victories, Neil Hodgson easily won the championship on the 999F03, now producing a claimed 189 horsepower at 12,500 rpm.

CONFUSED DIRECTIONS | 199

The 999R Fila was released to celebrate 200 World Superbike victories. In the background are the 999F03 racing machines of Neil Hodgson and Ruben Xaus. *Ducati*

BELOW RIGHT: Powered by the 999cc Testastretta, the 999R continued the tradition of limited-edition high-specification production homologation models.

The Ducati 998 continued to dominate the British Superbike Championship, but this year it was Shane Byrne, with twelve victories, who took the series on a recycled 998F02.

999 Fila, 999R, 999S, 999, 998, 749S, 749, 748

Heading the 2003 production lineup was the 999, joined shortly afterward by a similar 749. The 999 was initially available in three versions, the 999R, 999S, and 999, with the 749 as the 749S and 749, all powered by Testastretta engines. The earlier 998 and 748 continued for another year, but only in standard guises. During 2003, a limited-edition 999R Fila became available, this created to celebrate 200 World Superbike victories.

999, S, R, FILA, 749, S, 748 2003
Differing from the 998

Bore (mm)	90 (749)
Stroke (mm)	58.8 (749)
Displacement (cc)	748.14 (749)
Compression ratio	11.7:1 (749)
Power	103 horsepower at 10,000 rpm (749) 139 horsepower at 10,500 rpm (999R) 136 horsepower at 9,750 rpm (999S) 124 horsepower at 9,500 rpm (999)
Front suspension	43mm Öhlins (999R, 999S, 749R) 43mm Showa (999, 749S, 749)
Rear suspension	Showa (749S) Öhlins (999R, 999S, 749R) Boge (749)
Rear brake (mm)	240 disc
Rear tire	180/55x17 (749), 190/50x17 (999)
Wheelbase (mm)	1,420
Dry weight (kg)	199 (999, 999S, 749, 749S) 193 (999R)
Top speed (km/h)	240 (749) 250 (749S) 265 (999) 270 (999R, 999S)

Another product by Pierre Terblanche, the 999 retained many of Ducati's trademark features but was purposefully long, low, and narrow, to reduce the frontal area. The swingarm was now double-sided, the chassis designed to place more weight on the front wheel, and the 999R included racing-style radial Brembo front brake calipers for the first time on a production Ducati. Powering the 999 were engines very similar to those of the 2002 998. The 999R retained the shorter stroke 999cc engine, while the 999S shared the 998S motor, with the base 999 the lower output 998cc version.

New this year was the next evolution of the Testastretta, the 749. Replacing the 748, the 749 chassis was similar to the 999. As evidenced by its success in the World Superbike Championship, the 999 represented an improvement over the 998. The attention to weight distribution, the lower center of gravity, and the longer swingarm and wheelbase all contributed to improved stability under braking and acceleration, but buyers overlooked this functional superiority.

Similar to the 999, the 749 replaced the 748 for 2003. While improvements over the earlier 998/748, both the 999 and 749 met with resistance in the marketplace.

BELOW LEFT: Something midway between a street bike and trail bike, the 2003 Multistrada was a brilliant all-round motorcycle.

Multistrada, Supersport, Monster, Sport Touring

Also new for 2003 was Pierre Terblanche's Multistrada, or "many roads," designed to be equally at home in the mountains, city, or racetrack. Powering the Multistrada was an evolutionary 1,000cc Desmodue with dual spark plugs per cylinder, housed in a new tubular-steel frame with a single-sided swingarm. The traditional Supersport lineup was reengineered and repositioned in the marketplace. Now in three capacities, 620cc, 800cc, and 1,000cc, they included the 620 and 800 Sport, and 800 and 1000 Supersport. This year the 800 and 1000 engines replaced the 750 and 900 throughout the Supersport and Monster lineup, while the 620 was the same engine introduced in 2002 on the Monster 620. The

MULTISTRADA, SPORT TOURING 2003
Differing from 2002

Engine	90-degree V-twin four-stroke, air-cooled (Multistrada)
Bore (mm)	94 (Multistrada)
Stroke (mm)	71.5 (Multistrada)
Displacement (cc)	992 (Multistrada)
Compression ratio	10.5:1 (Multistrada)
Valve actuation	Belt-driven desmodromic single overhead camshaft (Multistrada)
Injection	Marelli 5.9M
Power	84 horsepower at 8,000 rpm (Multistrada)
Front suspension	43mm Showa (Multistrada)
Rear suspension	Showa (Multistrada)
Front brake (mm)	320 dual-disc (Multistrada)
Rear brake (mm)	245 disc (Multistrada)
Tires	120/70x17, 180/55x17 (Multistrada)
Wheelbase (mm)	1,462 (Multistrada)
Dry weight (kg)	188 (Multistrada) / 217 (ST4sABS)
Top speed (km/h)	208 (Multistrada)

SUPERSPORT 1000, 800, 620 2003
Differing from 2002 and the Multistrada

Bore (mm)	88 (800) 80 (620)
Stroke (mm)	66 (800) 61.5 (620)
Displacement (cc)	803 (800) 618 (620)
Power	85.5 horsepower at 7,750 rpm (1000) 74.5 horsepower at 8,250 rpm (800) 61 horsepower at 8,750 rpm (620)
Front suspension	43mm Showa 43mm Marzocchi (800, 620 Sport)
Rear suspension	Öhlins (1000)
Rear tire	180/55x17 (1000)
Wheelbase (mm)	1,395 (1000) 1,405 (800, 620)
Dry weight (kg)	185.2 (1000) 182.3 (800) 182 (620)
Top speed (km/h)	230 (1000) 210 (800) 195 (620)

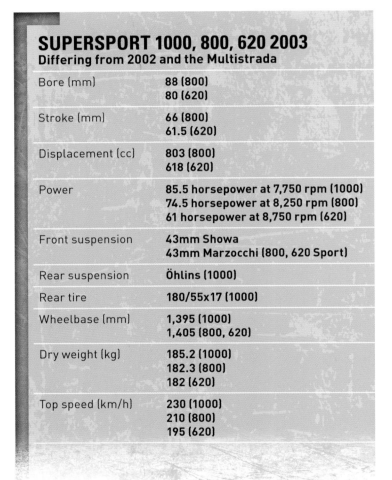

ABOVE LEFT: Heading the Supersport lineup for 2003 was the 1000DS, with the same engine as the Multistrada and 1000 Monster.

After ten years, for 2003 the Monster 1000 replaced highly successful M900. This is the Monster 1000 Dark, and the larger motor gave the Monster a new lease of life.

M1000, S, M800, S, 620 2003
Differing from 2002 and the Supersport

Compression ratio	10:1 (1000) 10.4:1 (800)
Power	73 horsepower at 8,250 rpm (800)
Rear tire	180/55x17 (1000)
Dry weight (kg)	179 (800)
Top speed (km/h)	220 (1000) 210 (800)

Multistrada 1000DS engine powered the 1000DS Supersport and Monster 1000 and 1000S, while the 800 was a development of the previous 750, now with a six-speed transmission. For this year, only a Monster 800S was offered alongside the Monster 800, as was the budget 620 and 800 Sport. New to the Sport Touring lineup was the ST4sABS.

2004

Despite the introduction of several new models, sales declined during 2004. The Supersport was particularly hard hit with production down 49.4 percent, to 1,426, while the Sport Touring and Multistrada also struggled. The 916cc ST4 and S4 were discontinued, and motorcycle

James Toseland on his way to winning the 2004 World Superbike Championship. *Ducati*

production declined to 36,560. During December 2004, the company also briefly laid off the workforce.

After surprising everyone with the Desmosedici in 2003, Ducati Corse produced an updated D16GP4 for the 2004 MotoGP season. Although the power was increased to 230 horsepower at 16,500 rpm, the D16GP4 didn't fulfill initial expectations, Capirossi finishing a lowly ninth in the World Championship. For the 2004 World Superbike Championship, Ducati signed James Toseland and Régis Laconi to ride the Fila 999F04s. The only official factory squad in the series, the 999s won ten races, Toseland narrowly winning the championship from Laconi. The power of the 999F04 was 189 horsepower at 12,500 rpm.

The Fila 999F04 factory World Superbike racer was an evolutionary development of the successful 2003 version. *Ducati*

999R, 999S, 999, 998 FE, Matrix, 749R, 749S, 749

Two final versions of the 998 were released this year: the 998FE (Final Edition) and 998 Matrix. The 998FE was similar to the earlier 998S, while the 998 Matrix was a 998, painted

One of the final 998s was the Matrix, as featured in the movie *The Matrix Reloaded*.

999, S, R, 749, S, R, DARK 2004
Differing from 2003

Bore (mm)	94 (749R)
Stroke (mm)	54 (749R)
Displacement (cc)	749.5 (749R)
Compression ratio	12.3:1 (749S); 12.7:1 (749R)
Injection	Marelli 5AM (749R)
Power	110 horsepower at 10,500 rpm (749S) 118 horsepower at 10,250 rpm (749R)
Dry weight (kg)	193 (749R) 192 (999R)

New for 2004 was the 749R, offering the highest specification yet for any production Ducati.

BELOW: For 2004, the ST3 replaced the ST2. It had a new fairing and three-valve engine.

ST3, ST4, ST4S 2004
Differing from 2003

Bore (mm)	94 (ST3)
Stroke (mm)	71.5 (ST3)
Displacement (cc)	992 (ST3)
Compression ratio	11.3:1 (ST3)
Valve actuation	Belt-driven desmodromic single overhead camshaft (three valves per cylinder) ST3
Power	102 horsepower at 8,750 rpm (ST3)
Dry weight (kg)	214 (ST3)
Top speed (km/h)	235 (ST3)

dark green as in the *Matrix* movie. This year, three versions of the 999 and 749 were available, including the standard, S, and R. While the 999R, 999S, and 999 were little changed, the 749 was significantly updated. Along with a new 749 Dark, the 749S motor was more powerful, and the 749R was the most technically advanced production bike ever offered by Ducati. Designed as a homologation model for World Supersport racing, where few modifications were allowed, this was ostensibly a racer with street equipment. Powered by a new short-stroke Testastretta with a slipper clutch, the fuel injection system was also a racing-type Marelli. While the frame was the same as the 749S, the aluminum swingarm was shared with the 2004 999R.

Sport Touring

Replacing the ST2 for 3004 was the ST3, powered by a new three-valve desmodromic engine (Desmotre). The crankcases, crankshaft, gearbox, and clutch were shared with the 1000DS, but the single overhead camshaft cylinder head was new. As on the 1000DS, ignition was by twin spark plugs. This engine combined the finest attributes of both the ST2 and ST4 and was eminently suited to the Sport Touring. The basic chassis was as on the previous ST2, and the new fairing and instrument panel was shared with the ST4s.

Monster, Supersport, and Multistrada

Even though the Supersport family was considerably revised the previous year, sales were below expectations. Only the 800 and 1000 were offered for 2004, both unchanged and with a full fairing. Also unchanged was the Multistrada. Celebrating ten years of production, and 130,000 manufactured, the Monster was considerably more successful than the Supersport and now included seven versions and four different engines. New this year was the S4R and 620. Replacing the Monster S4, the Monster 996cc S4R was the most powerful and best handling Monster yet, chassis updates including a tubular aluminum single-sided swingarm. The 620 featured an ATPC clutch to reduce back torque during deceleration, and while the 620 Dark retained the five-speed gearbox, other 620s, including a Monster Matrix 620 I.E. and 620S Capirossi (in honor of the factory MotoGP rider), now included a six-speed gearbox.

An exciting Monster for 2004 was the S4R, powered by the 996cc Desmoquattro engine.

Another variant of the 620 Monster for 2004 was the Capirossi, complete with the rider's No. 65. *Ducati*

MONSTER S4R, 1000, 800, 620 2004
Differing from the S4 and 2003

Bore (mm)	98 (S4R)
Displacement (cc)	996 (S4R)
Compression ratio	11.6:1 (S4R)
Power	113 horsepower at 8,750 rpm (S4R)
Gears	Six-speed (620 I.E, S)
Top speed (km/h)	240 (S4R)

2005

This year was another difficult one for Ducati. While there was some success in MotoGP, results in World Superbike were extremely disappointing, and motorcycle sales continued to fall, totaling 34,536 during 2005. Chairman Federico Minoli indicated a restructure and a change in direction, moving away from volume manufacture to smaller numbers of more expensive high-profit models. In MotoGP, Carlos Checa joined the Ducati Marlboro Team alongside Capirossi on the D16GP5. After the disappointment of trying to implement too many changes in the design of the D16GP4, Ducati returned to its traditional philosophy of "evolution not revolution," and toward the end of the season, Capirossi won two MotoGP races in succession (Japan and Malaysia). The claimed power was more than 230 horsepower at 16,500 rpm, and although in contention for second in the championship at one stage, a crash at Phillip Island ended Capirossi's hopes and he ended sixth overall.

Toseland and Laconi still rode the 999F05 in the World Superbike Championship, electronic updates increasing the power to 194 horsepower at 12,500 rpm. But results were disappointing, Toseland finishing fourth and Laconi sixth. In the British Superbike Championship, Gregorio Lavilla won the championship with seven race victories.

CLOCKWISE FROM TOP RIGHT:

Loris Capirossi on his way to victory in the 2005 Malaysian Grand Prix. *Ducati*

Toseland carried the No. 1 plate in the 2005 World Superbike Championship, but results were disappointing. *Ducati*

The D16GP5 was significantly improved over 2004. *Ducati*

999, S, R, 749, S, R, DARK 2005
Differing from 2004

Compression ratio	12.45:1 (999R)
Power	140 horsepower at 9,750 rpm (999)
	143 horsepower at 9,750 rpm (999S)
	150 horsepower at 9,750 rpm (999R)
	108 horsepower at 10,000 rpm (749)
	116 horsepower at 10,500 rpm (749S)
	121 horsepower at 10,500 rpm (749R)
Dry weight (kg)	188 (749)
	189 (749 Dark)
	186 (749S, 999, 999S)
	183.5 (749R)
	181 (999R)

999R, 999S, 999, 749R, 749S, 749

The 999 and 749 were significantly updated for 2005. All were more powerful, the fairing was redesigned, and the box-section swingarm was lighter. The higher specification 999R was designed to meet AMA regulations that required racing engines be almost identical to the corresponding production units. As before, the 999S slotted between the base 999 and limited-edition 999R, the 999S Öhlins front fork now with radial mount brake calipers.

Monster, Supersport, Sport Touring, and Multistrada

The Supersport's woes continued for 2005, with only the 1000DS now offered. Sport Touring sales were also down significantly this year, the range also continuing with only minor updates. Again, it was the Monster family that continued to sustain Ducati. Although more than a decade old, continual updates and range expansion ensured the

For 2005, the 999 was restyled, the fairing no longer with upper vents. *Ducati*

Monster's popularity. The Monster range doubled this year, with the S2R joining the S4R, along with three 1000s and three 620s. Powering the S2R was the two-valve 803cc engine, now with an APTC clutch. The S2R's style followed that of the S4R, with the right-side twin exhaust, small fairing, and single-sided aluminum swingarm. The Multistrada range was also expanded this year, including a more sporting Öhlins-equipped Multistrada 1000S and a pair of budget 620s, the 620 and 620 Dark.

MONSTER S2R, S4R, 1000, 620 2005
Differing from 2004

Bore (mm)	88 (S2R)
Stroke (mm)	66 (S2R)
Displacement (cc)	803 (S2R)
Compression ratio	10.5:1 (S2R)
Power	63 horsepower at 9,500 rpm (620)
	77 horsepower at 8,250 rpm (S2R)
	94 horsepower at 8,000 rpm (1000)
	117 horsepower at 8,750 rpm (S4R)
Front brake (mm)	300 twin-disc (S2R)
Dry weight (kg)	168 (620)
	173 (S2R)
	180 (1000)
	181 (S4R)

MULTISTRADA 1000, 620 2005
Differing from 2004

Bore (mm)	80 (620)
Stroke (mm)	61.5 (620)
Displacement (cc)	618 (620)
Compression ratio	10.5:1 (620)
	10:1 (1000)
Power	63 horsepower at 9,500 rpm (620)
	92 horsepower at 8,000 rpm (1000)
Front suspension	43mm Marzocchi (620)
	43mm Öhlins (1000SDS)
Rear suspension	Sachs (620)
	Öhlins (1000SDS)
Front brake (mm)	300 dual-disc (620)
Rear tire	160/60x17 (620)
Wheelbase (mm)	1,459 (620)
Dry weight (kg)	183 (620)
	196 (1000)

Powered by the 800cc Desmodue engine, the S2R was one of the most popular Monsters of 2005. *Ducati*

SPORT TOURING AND SUPERSPORT
Differing from 2004

Power	107 horsepower at 8,750 rpm (ST3)
	121 horsepower at 8,750 rpm (ST4S)
	95 horsepower at 7,750 rpm (1000SS)
Dry weight (kg)	201 (ST3)
	205 (ST4s)
	210 (ST4sABS)
	179.4 (1000SS)

2006

This year was significant for Ducati, marking the eightieth anniversary of the foundation of the company, the sixtieth anniversary of the start of motorcycle production, and the fiftieth anniversary of Taglioni's desmodromic valve system. Early in 2006, TPG sold its 30 percent stake in Ducati Motor to Investindustrial Holdings, but sales continued to decline, to 31,854. In September, Ducati sought ten weeks of temporary government-assisted layoffs. With a 999 replacement imminent, the most significant new releases this year were the SportClassic range and Monster S4RS, S2R1000, and 695.

Sete Gibernau joined Loris Capirossi on the D16GP6, and as this was the final year for 990cc, the GP6 was an evolution of the GP5. The weight was 148 kilograms and the claimed power in excess of 235 horsepower at 16,500 rpm. The season started brilliantly for Capirossi, with a victory in the opening race at Jerez, and at Mugello the bikes were raced in special livery to celebrate Ducati's three anniversaries. By midseason, Capirossi was leading the championship, but a crash at Catalunya ended his title hopes and he ultimately finished third overall. World Superbike Champion Troy Bayliss rode the 990cc D16GP6 in its final race at Valencia, leading from start to finish. He became the tenth rider to win races in Superbike and the premier GP class.

BELOW: The D16GP6 was an evolution of the GP5. *Ducati*

BELOW RIGHT: Loris Capirossi won three Grands Prix during 2006 and finished third in the championship. *Ducati*

Troy Bayliss returned to World Superbike in 2006 and easily won the championship. *Ducati*

Bayliss also made a return to the factory 999F06 in the World Superbike Championship, Lorenzo Lanzi partnering with him. Bayliss soon stamped his mark, winning eight races in succession in the first half of the season and finishing with twelve victories to easily win the championship. A triumph of evolution over revolution, the 999F06 produced 194 horsepower at 12,500 rpm and included the sophisticated Öhlins suspension and electronic traction control of the MotoGP D16GP6.

999 and 749

As replacement was imminent, the existing 999 and 749 continued ostensibly unchanged for 2006. Heading the lineup was the 999R Xerox, with the colors and graphics of the World Superbike team and racing-type rear Öhlins shock absorber.

Supersport, Sport Touring, and Multistrada

Although sales were virtually nonexistent, the 1000DS Supersport continued for one more year. The Multistrada was also unchanged this year while the Sport Touring range was rationalized to include the ST3 and ST3sABS, this replacing the ST4sABS. The ST3sABS included the ST4s's ABS anti-lock braking system and higher specification suspension.

The limited-edition 999R Xerox was painted in the same livery as Toseland's 999F05. *Ducati*

Monster

With sales now totaling 170,000, the Monster remained the backbone of Ducati's range and the lineup focused on the expansion of the SR series, with two S2Rs, 800 and 1,000cc, and a new model, the S4RS Testastretta. The most powerful Monster yet, the S4RS featured the 998cc Testastretta motor, a stiffened frame, and an Öhlins front fork with radial front brake calipers. The 996cc Monster S4R continued as before, and the S2R1000 replaced the previous Monster 1000.

ST3, ST3SABS 2006 Differing from 2005

Rear suspension	Öhlins (ST3sABS)
Dry weight (kg)	204 (ST3sABS)

The ST3sABS replaced the ST4sABS for 2006. The wheels were five-spoke, and the front fork included TiN-coated fork stanchions. *Ducati*

With its Öhlins suspension and 998cc Testastretta engine, the Monster S4RS was a naked Superbike. *Ducati*

MONSTER S4RS, S2R, S4R, 620 2006
Differing from 2005

Bore (mm)	100 (S4RS)
Stroke (mm)	63.5 (S4RS)
Displacement (cc)	998 (S4RS)
Compression ratio	11.4:1 (S4RS)
Power	95 horsepower at 8,000 rpm (S2R1000) 130 horsepower at 9,500 rpm (S4RS)
Front suspension	43mm Öhlins (S4RS)
Rear suspension	Öhlins (S4RS)
Front brake (mm)	320 twin-disc (S2R1000, S4RS)
Dry weight (kg)	178 (S2R1000) 177 (S4RS)

SportClassic

Headed by the limited-edition Paul Smart 1000, the SportClassic series was a modern interpretation of the great trio of 1970s 750 round-case Ducatis. The Ducati 750 Sport and Super Sport of 1972–1974 were long and low, with curving lines and distinctive colors, Pierre Terblanche seeking to emulate this style with an air-cooled 1000DS engine and exposed tubular-steel frame. The 1000DS engine was shared with the Multistrada, Supersport, and Monster S2R. Only 2,000 examples of the Paul Smart 1000 Limited Edition were produced, and as a contemporary interpretation of the iconic 750 Super Sport, it included high-quality

The Paul Smart 1000 Limited Edition was a modern interpretation of the classic 1974 750 Super Sport, here in the background. *Ducati*

chassis components. The Sport 1000 was similar in style to the original 1973 750 Sport, an elemental sportbike without a fairing. Completing the SportClassic lineup was the GT1000, this reflecting the line and the shape of the 1971 750GT, a basic two-seater with exposed engine and single headlight, much like a modern naked motorcycle.

SPORTCLASSIC PAUL SMART, SPORT, GT 1000 2006
Differing from the 1000DS Supersport

Power	92 horsepower at 8,000 rpm
Front suspension	43mm Öhlins (PS1000) 43mm Marzocchi (Sport, GT)
Rear suspension	Öhlins (PS1000) Sachs (Sport, GT)
Wheelbase (mm)	1,425
Dry weight (kg)	181 (PS1000) 179 (Sport) 185 (GT)

The Sport 1000 was an updated version of the 1973 750 Sport. *Ducati*

RIGHT: Completing the retro-750 theme was the GT1000, with twin-shock rear suspension and dual mufflers. *Ducati*

CHAPTER 9
GRAND PRIX SUCCESS
MOTOGP, DESMOSEDICI, 1098, 2007–2011

Casey Stoner completely dominated the 2007 MotoGP season, easily winning the championship. *Ducati*

Ducati's 2007 winning team. Stoner is on the right and Capirossi on the left. *Ducati*

After a few difficult years, things began to look up for Ducati in 2007. Casey Stoner won the MotoGP World Championship on the new 800cc D16GP7, and the new 1098 replaced the 999. Over the next few years, more significant models were released, including the Hypermotard, production Desmosedici, Streetfighter, groundbreaking Multistrada 1200, and radical Diavel.

2007

Buoyed by the unexpected success of the new 1098 and Hypermotard, motorcycle production increased to 39,575 during 2007. Gabriele Del Torchio was appointed the new chief executive officer, and in MotoGP twenty-one-year-old Australian Casey Stoner joined the veteran Capirossi, now in his fifth season on the Desmosedici. The desmodromic 800cc V-four D16GP7 revved beyond 20,000 rpm and produced around 220 horsepower. The chassis still included a steel frame but with a carbon-fiber swingarm and rear subframe. The Öhlins suspension was similar to the D16GP6, the weight 148 kilograms, and Stoner blitzed the championship with a total of ten race wins.

Now powered by an 800cc Desmosedici, the GP7 looked quite similar to the GP6. *Ducati*

With the new 1098 already in production, the World Superbike Team marked time with the 999F07. Now at the peak of its development, the 999cc engine was still much as it had been over the past three seasons and produced 194 horsepower at 12,500 rpm. Bayliss and Lanzi both struggled, Bayliss ultimately finishing fourth in the championship with seven race wins.

1098, 1098S

With the announcement that World Superbike regulations would soon allow 1,200cc twins, Ducati released the 1098 for 2007. In a departure from Terblanche's radical and unloved 999, and designed in-house by a team led by Gianni Fabbro, the 1098 emulated the style of the earlier 916. Striving for more power and less weight was Ducati's aim with the 1098, and as was quite usual with Ducati, the model designation was a misnomer. The Testastretta Evoluzione displaced 1,099cc. After the unpopularity of the 999's single muffler, the 1098 included a pair of 916-style under-seat mufflers. The new engine was housed in a trademark steel trellis

RIGHT: The 1098 and 1098S replaced the 999 for 2007 and both were immediately successful. *Ducati*

BELOW: Also available in 2007 was a limited-edition 1098S Tricolore. *Ducati*

frame, with a return to a single-sided swingarm. A limited-edition 1098S Tricolore was also available, and although the 1098 was ineligible for World Superbike racing in 2007, its excellence was clearly demonstrated in the FIM Superstock World Cup, Niccolò Canepa narrowly taking the title.

Hypermotard 1100, 1100S

Also introduced for 2007 was a new family, the Hypermotard. Terblanche's final project for Ducati, the Hypermotard was powered by an 1,100cc evolution of the two-valve, air-cooled, desmodromic dual-spark engine. The steel-trellis frame was new and also included a single-sided swingarm. Other than improved wheels, brakes, and suspension, the Hypermotard S was similar to the standard Hypermotard.

1098, 1098S 2007
Differing from the 999

Bore (mm)	104
Stroke (mm)	64.7
Displacement (cc)	1,099
Compression ratio	12.5:1
Power	160 horsepower at 9,750 rpm
Front suspension	43mm Öhlins (1098S) 43mm Showa (1098)
Rear suspension	Öhlins (1098S) Showa (1098)
Front brake (mm)	330 dual-disc
Rear brake (mm)	245 disc
Rear wheel	6.00x17
Rear tire	190/55x17
Wheelbase (mm)	1,430
Dry weight (kg)	171 (1098S) 173 (1098)

Factory rider Ruben Xaus demonstrating the Hypermotard's abilities. *Ducati*

HYPERMOTARD 1100, 1100S 2007

Engine	Twin-cylinder four-stroke, air-cooled
Bore (mm)	98
Stroke (mm)	71.5
Displacement (cc)	1,078
Compression ratio	10.5:1
Valve actuation	Belt-driven desmodromic single overhead camshaft, 2 valves per cylinder
Power	90 horsepower at 7,750 rpm
Front suspension	50mm Marzocchi
Rear suspension	Öhlins (Hypermotard S) Sachs (Hypermotard)
Front brake (mm)	305 dual-disc
Rear brake (mm)	245 disc
Wheels	3.50x17 and 5.50x17
Tires	120/70x17, 180/55x17
Wheelbase (mm)	1,455
Dry weight (kg)	177 (Hypermotard S) 179 (Hypermotard)

Apart from the new 1,100cc engine, the Multistrada 1100 was little changed from the 1,000cc version. The 1100S had an Öhlins suspension. *Ducati*

MULTISTRADA 1100, 1100S 2007
Differing from 2006 and the Hypermotard

Power	95 horsepower at 7,750 rpm
Front suspension	43mm Marzocchi (1100)
Rear suspension	Sachs (1100)

Multistrada 1100, 1100S

The Multistrada was also updated with the new 1,100cc engine for 2007. Unlike the Hypermotard, the Multistrada had a wet clutch and the engine produced slightly more power. The basic style was unchanged from the 1000, with the same chassis, and the 1100S still included fully adjustable Öhlins suspension.

Monster, SportClassic, Sport Touring

Updates to the Monster range included a new S4R, now with a Testastretta engine, and entry-level 695. The existing S4RS and S2R were largely unchanged. The SportClassic also continued for 2007, the Sport 1000 as a traditional Monoposto or Biposto, both with twin Sachs shock absorbers and dual mufflers. As the Paul Smart 1000 finished in 2006, a Sport 1000S with half fairing was available this year, while the GT1000 continued much as before, as did the ST3 and ST3ABS.

With the Paul Smart 1000 finished, the half-faired 1000S was available for 2007. *Ducati*

The Monster S4R gained the 998cc Testastretta engine for 2007. *Ducati*

MONSTER S4RS, S4R, S2R, 695, SPORT 1000, 1000S, GT1000 2007
Differing from 2006

Bore (mm)	88 (695)
	100 (S4R)
Stroke (mm)	57.2 (695)
	63.5 (S4R)
Displacement (cc)	695 (695)
	998 (S4R)
Compression ratio	10.5:1 (695)
	11.4:1 (S4R)
Power	73 horsepower at 8,500 rpm (695)
	130 horsepower at 9,500 rpm (S4RS)
Front suspension	43mm Marzocchi (695)
	43mm Showa (S4R)
Rear suspension	Sachs (695, S4R)
Front brake (mm)	300 twin-disc (695)
Dry weight (kg)	168 (695)
	177 (S4R)
	181 (Sport 1000S)

2008

In the wake of the exceptional success on the racetrack and in the showroom during 2007, new 2008 models included the Desmosedici RR, 1098R, 848, and 696 Monster. And despite a general market decline, motorcycle production increased nearly 20 percent, to 36,132 by the end of September 2008.

Marco Melandri replaced Capirossi alongside Stoner in the MotoGP team on the D16GP8, but early results were disappointing. Ultimately, Stoner won six races to finish second in the championship. In the World Superbike Championship, the 1098F08 was based on the newly homologated 1098R, the power of the 1,198cc twin a claimed 198 horsepower at 11,000 rpm. Michel Fabrizio joined Bayliss in the Xerox Ducati team, and with eleven race wins, Bayliss easily won his third World Superbike Championship. Bayliss retired as one of Ducati's most successful riders, and the 1098RS was also successful in the British Superbike Championship, Shane Byrne winning the title with eleven victories.

The GP8 was an evolution of the GP7, but was more difficult to ride. *Ducati*

Desmosedici RR

First shown during 2006, and initially scheduled for 2007 production of only 400 units, the production Desmosedici

LEFT: Although he failed to win the 2008 MotoGP World Championship, Casey Stoner was still the only rider who could come to terms with the tricky Desmosedici. *Ducati*

BELOW LEFT: The 1098F08 World Superbike racer was immediately successful. *Ducati*

BELOW: Troy Bayliss easily won the 2008 World Superbike Championship. For his final race at Portmão in Portugal, his 1098F08 had a special color scheme and he comfortably won both races. *Ducati*

was finally available in 2008. Based on the 990cc GP6, the 16-valve desmodromic V-four included gear-driven double overhead camshafts and a six-speed cassette-type gearbox. The welded tubular-steel frame with aluminum swingarm shared its geometry with the GP6, while the Öhlins suspension and forged Marchesini wheels were also similar. Two versions were available: the Rosso GP and Team Version with factory Corse livery. Fifteen hundred were eventually built, and it was the only genuine MotoGP replica available to the general public. Even at the $72,500 list price, the D16RR was a bargain, and it's doubtful if it was a profitable exercise for Ducati.

BELOW: The production Desmosedici RR was closely modeled on the racing GP6. *Ducati*

1098R, 1098S, 1098, 848

Joining the successful 1098 and 1098S for 2008 was the homologation 1098R and smaller 848. The change in World Superbike regulations now required the "R" to be closer to the factory Superbikes in specification, and the 1098R was Ducati's most exotic twin yet. With a larger bore and longer stroke, larger valves, higher lift camshafts, and elliptical

GRAND PRIX SUCCESS | 217

Virtually a Grand Prix racer with lights, the D16RR amazingly was still street legal. *Ducati*

throttle bodies, the 1098R produced 186 horsepower with the optional race kit, and it was good enough for Brendan Roberts to win the FIM Superstock World Cup. With a chassis similar to the 1098, the 848 replaced the 749.

Monster, Hypermotard, Multistrada, SportClassic

By 2008, the Monster lineup was looking dated and in need of a makeover, and the most radical update yet appeared with the 696. The 696 was a very clever design that managed to uphold the Monster philosophy, emphasizing the minimalist approach but in a new modern format. The 696 engine was an evolution of the previous 695 version, with the same bore and stroke but with cylinder heads similar to the new 1,100cc dual-spark unit. This revised motor was mounted in a totally new steel frame, with improved suspension and brakes.

There were few changes to the other Monsters for 2008. The S2R continued as 1,000cc only, and a Special Edition Tricolore S4RS now headed the Monster family lineup. This was a final tribute to the traditional Monster line that first appeared in 1992. The Hypermotard, Multistrada, and SportClassic continued largely unchanged.

DESMOSEDICI RR 2008

Engine	Four-cylinder four-stroke, liquid-cooled
Bore (mm)	86
Stroke (mm)	42.56
Displacement (cc)	989
Compression ratio	13.5:1
Valve actuation	Gear-driven desmodromic double overhead camshafts, four valves per cylinder
Power	200 horsepower at 13,800 rpm
Gears	Six-speed
Front suspension	43mm Öhlins
Rear suspension	Öhlins
Front brake (mm)	330 dual-disc
Rear brake (mm)	240 disc
Tires	120/70x17, 200/55x16
Wheelbase (mm)	1,430
Dry weight (kg)	171

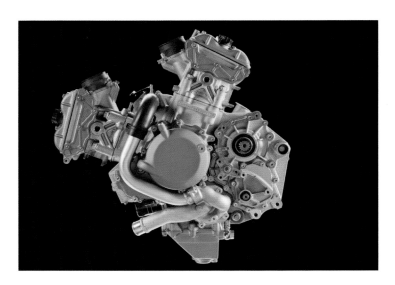

The Desmosedici 989cc V-four included gear-driven double overhead camshafts. *Ducati*

Displacing 1,099cc, the 1098R was the homologation model for World Superbike racing. *Ducati*

A smaller version of the 1098, the 848 joined the lineup for 2008. *Ducati*

1098R, S, 1098, 848 2008
Differing from 2007 and the 1098

Bore (mm)	106 (1098R) 94 (848)
Stroke (mm)	67.9 (1098R) 61.2 (848)
Displacement (cc)	1,198.4 (1098R) 848 (848)
Compression ratio	12.8:1 (1098R) 12:1 (848)
Power	180 horsepower at 9,750 rpm (1098R) 134 horsepower at 10,000 rpm (848)
Front suspension	43mm Öhlins (1098R) 43mm Showa (848)
Rear suspension	Öhlins (1098R) Showa (848)
Front brake (mm)	320 dual-disc (848)
Rear wheel	5.50x17 (848)
Rear tire	180/55x17 (848)
Dry weight (kg)	165 (1098R) 168 (848)

MONSTER 696, S4RS, S4R, S2R 695 2008
Differing from 2007 and the 695

Compression ratio	10.7:1 (696)
Injection	Siemens (696)
Power	80 horsepower at 9,000 rpm (696)
Front suspension	43mm Showa (696)
Front brake (mm)	320 twin-disc (696)
Wheelbase (mm)	1,450 (696)
Dry weight (kg)	161 (696)

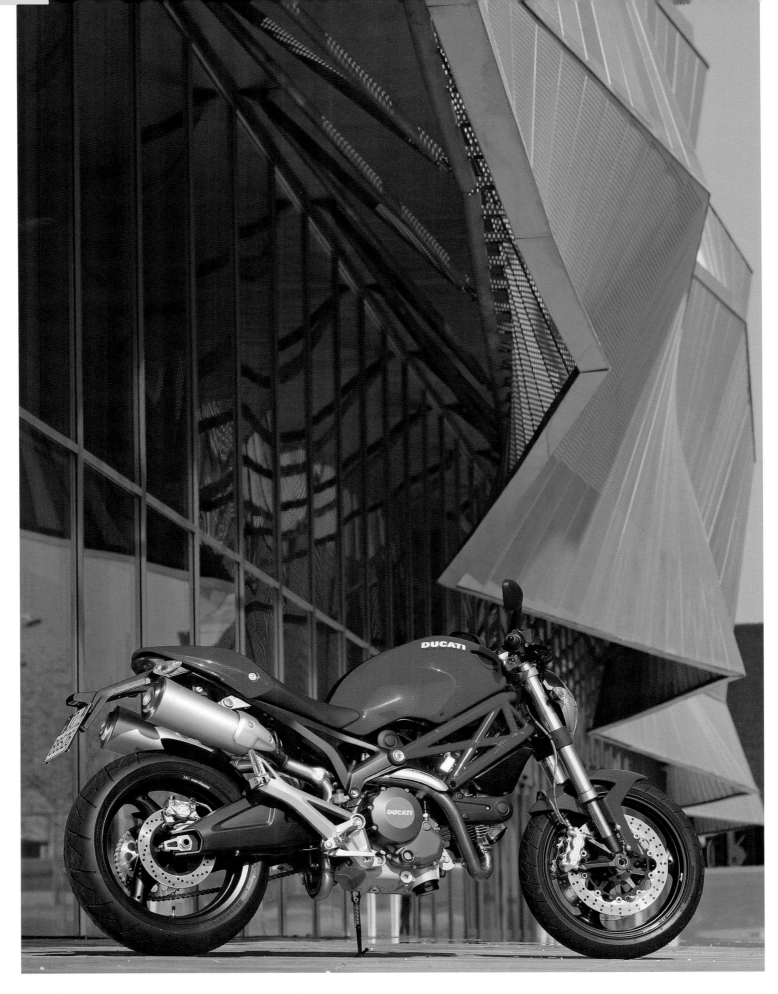

The 2008 696 rejuvenated the Monster range. *Ducati*

The Special Edition Tricolore S4RS headed the 2008 Monster lineup. *Ducati*

2009

On the wave of success in the World Superbike Championship, the 1098 became the 1198 this year, and the Streetfighter and Monster 1100 were introduced. But in the wake of the financial crisis across Europe and America, sales went into decline, totaling only 34,605 during 2009. In MotoGP, 2006 World Champion Nicky Hayden joined Casey Stoner, the D16GP9 now including a carbon-fiber chassis and swingarm. Although the season began well with Stoner winning at Qatar and Mugello, the 2009 season unfolded as one to forget for Ducati. Hayden struggled to get up to speed, and Stoner was ravished by a mystery illness that saw him slump midseason and miss three races. Stoner resurrected some of his form with blistering wins at Phillip Island and Sepang to finish fourth in the World Championship.

Following Troy Bayliss's retirement, veteran World Superbike rider Noriyuki Haga was signed to ride the 1198F09 alongside Fabrizio. The 1198F09 was largely unchanged, and after a strong start to the season, some crucial crashes resulted in Haga narrowly losing the championship.

Despite winning several races, a mystery illness destroyed Stoner's chances of taking the championship. *Ducati*

LEFT: The D16GP9 had a carbon-fiber frame and swingarm. *Ducati*

ABOVE: Nicky Hayden joined Stoner on the GP9 but struggled to find form. *Ducati*

BELOW: The 2009 1098F09 World Superbike racer was similar to the 2008 model. *Ducati*

1198, 1198S, 1098R, 1098R Bayliss, 848

Heading the Superbike lineup this year was a limited-edition 1098R Bayliss, and although the 1098 was extremely successful, the 1198 replaced it. Both the 1198 and 1198S now shared the engine dimensions of the 1098R, along with larger valves and racing-type rockers. The power was increased, and the 1198S now included electronic traction control. To celebrate the retirement of three-time World Superbike Champion Troy Bayliss, 500 1098R Bayliss Limited Editions were produced. The specification was as for the 1098R, but the Bayliss Limited Edition had a special Aldo Drudi—designed color scheme. The regular 1098R and 848 continued unchanged. Although the 1198 was unsuccessful in the World Superbike Championship, Belgian rider Xavier Simeon won the Superstock 1000 FIM Cup on the Ducati Xerox Junior Team 1198S. Simeon won five races and comfortably won the championship.

The 1198 replaced the 1098 for 2009. Here is the 1198S in the foreground with an 1198 behind. *Ducati*

BELOW: The 1098R Bayliss was painted in colors to replicate those used in Bayliss's final World Superbike race. *Ducati*

1198S, 1198, 1098R, BAYLISS, 848 2009
Differing from 2008 and the 1098

Bore (mm)	106 (1198/S)
Stroke (mm)	67.9 (1198/S)
Displacement (cc)	1,198.4 (1198/S)
Compression ratio	12.7:1 (1198/S)
Power	170 horsepower at 9,750 rpm (1198/S)
Dry weight (kg)	171 (1198) 169 (1198S)

Powered by the 1098 Testastretta Evoluzione, the Streetfighter took the powerful naked bike concept to a new level. The Streetfighter S had high-specification Öhlins suspension. *Ducati*

Streetfighter, Streetfighter S

The Streetfighter joined the lineup for 2009, the product of in-house designer Damien Basset. This effectively replaced the Monster S4R and S4RS and was ostensibly a naked 1098 Superbike. Much of the Streetfighter was new, the Testastretta engine combining 1098 cylinder heads with 1198 crankcases. With more rake and a longer swingarm, the frame was also specific to the Streetfighter, and the S version received Öhlins suspension, traction control, and data analysis.

Monster, Hypermotard, Multistrada, SportClassic

Another significant new model for 2009 was the Monster 1100. Following the release of the new-generation 696, this appeared in the guise of the 1100 and 1100S. Apart from the 696, all other Monsters were discontinued this year. Central to the Monster 1100's design was an updated 1,100cc air-cooled dual-spark Desmodue twin. The steel-trellis frame was developed in conjunction with Ducati Corse and shared the Desmosedici short frame concept and the same tube diameter and thickness as the 1098R. The general style followed that of the 696, with the distinctive triple-parabola headlight, the 1100S also including Öhlins suspension. Apart from new colors choices, there were no changes to the Hypermotard 1100 or Multistrada 1100 for 2009, but the Hypermotard 1100S was updated with a new front fork. The SportClassic included the 1000S, GT1000, and the new GT 1000 Touring this year.

STREETFIGHTER, S 2009
Differing from the 1098 and 1098S

Power	155 horsepower at 9,750 rpm
Wheelbase (mm)	1,474
Dry weight (kg)	169 167 (S)

MONSTER 1100, S, 696, HYPERMOTARD 2009 Differing from 2008

Bore (mm)	98 (M1100)
Stroke (mm)	71.5 (M1100)
Displacement (cc)	1,078 (M1100)
Compression ratio	10.7:1 (M1100)
Injection	Siemens (M1100)
Power	95 horsepower at 7,500 rpm (M1100)
Front suspension	43mm Öhlins (M1100S) 43mm Showa (M1100) 48mm Kayaba (Hypermotard S)
Rear suspension	Öhlins (M1100S) Sachs (M1100)
Wheelbase (mm)	1,450 (M1100)
Dry weight (kg)	168 (M1100S) 169 (M1100)

Continuing the 696 Monster's theme for 2009 was the Monster 1100. *Ducati*

Stoner struggled to overcome the GP10's front-end problems during 2010. *Ducati*

2010

Despite a difficult time for the motorcycle industry in general, Ducati continued to release new and updated models, sales increasing slightly to 36,050. There were now sixteen models in six families: five Superbikes, two Streetfighters, two Multistrada, three Monsters, and one SportClassic. The Hypermotard gained an entry-level model, the 796, while the Hypermotard S evolved into the more exclusive SP. But the most important new release this year was the Multistrada 1200, a motorcycle that expanded many design parameters in the sport touring sector.

Circumstances were against Ducati in MotoGP this year. Casey Stoner had proven to be the only rider to come to terms with the difficult 800cc Desmosedici, but during the season he became unsettled and began looking for an alternative ride. Although Nicky Hayden still struggled to adapt to the D16GP10, changes in riding position and electronics saw an improvement in his performance. New this year was a 48mm Öhlins front fork, but the D16GP10 still suffered front-end problems. Stoner crashed out while in the lead at Qatar, and it wasn't until midseason that he finished on the podium. Three victories toward the end of the season saw him finish fourth overall, with Hayden seventh.

Ducati had a torrid time in the 2010 World Superbike Championship. The riders, Haga and Fabrizio, were unchanged, and while the 1198F10 was little changed, the Ducatis were penalized by air restriction and weight regulations. The power was 200 horsepower at 11,000 rpm, and after beginning the season weighing 168 kilograms, the Ducatis received a weight reduction to 162 kilograms after disappointing results. With three victories, Haga managed sixth overall, the least successful season ever for the factory team in the World Superbike Championship.

ABOVE: The GP10 was an evolution of the GP9. *Ducati*

BELOW: Noriyuki Haga had a torrid time on the 1198F10 in the 2010 World Superbike Championship. Poor results saw the factory withdraw from the series. *Ducati*

ABOVE: The 1198R Corse Special Edition both featured a factory team-style aluminum gas tank, Ducati traction control, race kit, and new Ducati Corse colors. *Ducati*

RIGHT: Distinguished by a personally autographed fuel tank, the Nicky Hayden 848 was produced exclusively for the US market. *Ducati*

1198R, S, 1198, CORSE, 848 2010
Differing from 2009

Dry weight (kg)	168 (1198S Corse)
	164 (1198R Corse)

848, 1198, 1198S, 1198R, Corse, Nicky Hayden Special Editions

To celebrate winning the 2009 World Superbike and Superstock Manufacturers' titles, Ducati presented two Corse special editions, the 1198S and 1198R Corse Special Editions. Released as an early 2010 model year example at the 2009 United States Grand Prix at Laguna Seca, the US-only Nicky Hayden 848 was noted for its American-inspired color scheme, but the rest of the 1198 and 848 range was ostensibly unchanged.

Multistrada 1200

Following its 2003 release, the Multistrada established a new motorcycle genre, one that combined sports performance, trail blazing, and unrivalled versatility. After beginning life with the air-cooled 1,000cc Dual Spark engine, and receiving the 1,100cc version in 2006, for 2010 the Multistrada was completely redesigned. Hailed as four bikes in one, the Multistrada 1200 also established new technological boundaries.

Powering the Multistrada 1200 was the 1198 Testastretta Evoluzione, with valve overlap reduced to 11 degrees to minimize emissions and improve fuel consumption. The trellis frame was signature Ducati, with large diameter, light-gauge tubing with two central cast-aluminum sections and a single-sided aluminum swingarm. The suspension was also new, "S" versions equipped with Ducati Electronic Suspension (DES) that allowed spring preload and rebound and compression damping adjustment to be electronically controlled via the instrument panel. A Bosch-Brembo ABS system was optional on the standard version and standard on the S. The most revolutionary feature was the choice of four riding modes: Sport, Touring, Urban, and Enduro, each offering preset power, suspension setup, and traction control for specific conditions.

With its advanced electronic suspension and selection of riding modes, the Multistrada 1200S was one of the most versatile motorcycles available in 2010. *Ducati*

Hypermotard 796, 1100 EVO, 1100 EVO SP

For 2010, the Hypermotard's less weight and more power philosophy was expanded and the range extended to include the entry-level 796 and higher specification 1100 EVO and SP. Powering the Hypermotard 796 was a new engine, not simply a stroked 696. A new frame provided a 20mm lower seat height than the 1100. Joining the 796 was a lighter and more powerful 1100 EVO (evolution) and SP Hypermotard. More extreme, the SP provided increased suspension travel and more ground clearance.

MULTISTRADA 1200, S 2010 Differing from the 1100MS 2009	
Engine	Twin-cylinder four-stroke, liquid-cooled
Bore (mm)	106
Stroke (mm)	67.9
Displacement (cc)	1,198.4
Compression ratio	11.5:1
Valve actuation	Belt-driven desmodromic double overhead camshafts, four valves per cylinder
Injection	Mitsubishi
Power	150 horsepower at 9,250 rpm
Front suspension	48mm Öhlins (S) 50mm Marzocchi (MS)
Rear suspension	Öhlins (S) Sachs (MS)
Rear wheel	6.00x17
Rear tire	190/55x17
Wheelbase (mm)	1,530
Dry weight (kg)	192 (S) 189 (MS)

Offering increased ground clearance, 1198 Brembo Monobloc brake calipers, and forged Marchesini wheels, the 1100 EVO SP was the most radical Hypermotard yet. *Ducati*

HYPERMOTARD 796, 1100 EVO, SP 2010
Differing from 2009

Bore (mm)	88 (796)
Stroke (mm)	66 (796)
Displacement (cc)	803 (796)
Compression ratio	11:1 (796) 11.3:1 (EVO)
Injection	Siemens (EVO, 796)
Power	81 horsepower at 8,000 rpm (796) 95 horsepower at 7500 rpm (EVO)
Front suspension	43mm Marzocchi (796)
Rear suspension	Öhlins (SP)
Wheelbase (mm)	1,465 (SP)
Dry weight (kg)	167 (796) 172 (EVO) 171 (SP)

MONSTER 1100, S, 696, ABS 2010
Differing from 2009

Dry weight (kg)	163 (696ABS) 171 (1100ABS) 170 (1100SABS)

Streetfighter, Monster, SportClassic

The two-model Streetfighter range was unchanged for 2010, as was the Monster in 696 and 1,100cc versions. New for all Monsters this year was the option of ABS and Monster Art kits, allowing a change to the fuel tank, fairing, front fender, and rear seat cover colors. Now in its final year, the only SportClassic was the GT1000.

2011

Despite a shrinking international market, 42,233 sales during 2011 made this year the most successful in Ducati's history. Sales were particularly strong in North America, turnover increasing more than 60 percent. Under the control of International Motorcycles S.p.A., a subsidiary of Investindustrial, a new factory was established in Thailand, and the number of employees increased to 1,135. But the big news for 2011 was the signing of megastar nine-time World Champion Valentino Rossi in a two-year agreement to ride for the MotoGP Ducati Marlboro Team. Nicky Hayden was also signed for an additional two years.

Unfortunately, what should have been the Italian dream team turned into a nightmare. In the final year for 800cc, the D16GP11 was initially structurally little changed from the GP9/GP10 but during the season incorporated some updates intended for the new 1,000cc GP12 and eventually an aluminum beam frame. The 800cc V-four Desmosedici produced around 225 horsepower, revving to 18,500 rpm, but Rossi's inability to adapt to the D16GP10 saw wins drying up

Valentino Rossi's D16GP11 was initially little changed from 2010. *Ducati*

completely, his best result a solitary podium at Le Mans in fortuitous circumstances. After crashing out of the last four races, he finished seventh, just ahead of Hayden in eighth.

Ducati withdrew its factory World Superbike team for 2011, and although the 1198R was due for replacement, Ducati provided the 1198RS11 customer racer to Team Althea for the thirty-eight-year-old veteran Carlos Checa. New camshafts and updated software resulted

LEFT: Rossi struggled to come to terms with the GP11 and had an extremely disappointing season. *Ducati*

BELOW: Carlos Checa surprised the pundits by easily winning the 2011 World Superbike Championship on the 1198RS11. *Ducati*

in slightly more power, now around 200 horsepower at 11,000 rpm. While on paper the aging twin looked outclassed, with fifteen race wins Checa dominated the World Superbike Championship and easily took the title.

1198SP, 1198, 848 EVO

With replacement imminent, the Superbike lineup comprised only three models for 2011. The 1198SP replaced the 1198S and 1198R, and it combined the standard 170-horsepower 1198 engine in the 1198R chassis with Öhlins suspension. Both 1198s gained traction control, data analysis, and a quick shift system as standard, and with four victories in the FIM Superstock 1000 FIM Cup on the Team Althea Racing Ducati 1198, Davide Giugliano won the series.

While the engine specification of the two 1198s was unchanged, with new camshafts and elliptical throttle bodies, the 848 EVO Testastretta Evoluzione received a horsepower boost. In March, Jason DiSalvo took the Team Latus Motors Racing 848 EVO to a hard-fought victory in the Daytona 200, Ducati's first win at the legendary circuit for thirty-four years.

1198SP, 1198, 848 EVO 2011
Differing from 2010

Compression ratio	13.2:1 (848 EVO)
Power	140 horsepower at 10,500 rpm (848 EVO)
Dry weight (kg)	168 (1198SP)

GRAND PRIX SUCCESS

A huge tire dominated the Diavel's rear aspect. *Ducati*

DIAVEL, CARBON 2011
Differing from the Multistrada 1200

Power	162 horsepower at 9,500 rpm
Front suspension	50mm Marzocchi
Rear suspension	Sachs
Rear brake (mm)	265
Rear wheel	8.00x17
Rear tire	240/45x17
Wheelbase (mm)	1,590
Dry weight (kg)	210 (Diavel) 207 (Carbon)

Diavel, Diavel Carbon

Also introduced during 2011 was the Diavel, a custom muscle bike in the mold of the Yamaha V-Max or Harley-Davidson Night Rod, but lighter and more agile. The Diavel combined the features of an imposing comfortable cruiser with those of a powerful sportbike.

Derived from the Multistrada 1200, the Diavel included the Testastretta 11° engine, ride-by-wire throttle, traction control, riding modes, and ABS. Specific to the Diavel was an intricate 2-into-1-into-2 exhaust system and tank/airbox with large air scoops around the front fork. Also new was a long, single-sided swingarm and wide rear rim for the massive rear tire. Along with optional luggage, windshield, and backrests, the Diavel was available in two versions: the Diavel and Diavel Carbon. The Carbon featured carbon-fiber bodywork, a performance exhaust, and lighter Marchesini wheels.

Monster, Streetfighter, Multistrada, Hypermotard

The Monster range expanded for 2011 with the early release 796 followed by the Monster 1100 EVO. A further range of colors was available, with Monster Art extended to thirteen with the new collection, Logomania. These were a tribute to historic Ducati logos and color schemes spanning more than fifty years, including Valentino Rossi and Nicky Hayden GP Replicas.

For the Monster 1100 EVO, the power was increased and the under-seat exhaust system replaced by a curving side-mounted system similar to the Diavel. The riding position was more relaxed, with a new seat and 20mm bar riser, while electronic additions included ABS and traction control. The Monster 1100, 1100S, and 696 were ostensibly unchanged.

Also available for 2011 was the Monster 796, with the 796 Hypermotard engine, the 796's style differing slightly to both the 1100 and 696. The frame, with single-sided swingarm, and the suspension brakes and wheels were shared with the 1100, but the riding position was revised with a lower seat and higher handlebars. The Streetfighter, Multistrada, and Hypermotard received only cosmetic updates for 2011, with a Multistrada 1200S Pikes Peak Special Edition produced to celebrate Greg Tracy's victory in the 2010 Pikes Peak International Hill Climb race in Colorado. Carlin Dunne repeated Tracy's victory in the 2011 event.

MONSTER 1100 EVO, 1100S, 1100, 796, 696 2011
Differing from 2010 and the 796 HM

Compression ratio	11.3:1 (EVO)
Power	100 horsepower at 7,500 rpm (EVO) 87 horsepower at 8,250 rpm (796)
Front suspension	43mm Marzocchi (EVO, 796, 696)
Rear suspension	Sachs (EVO, 796)
Dry weight (kg)	167 (796) 169 (EVO, 796ABS)

With its 100 horsepower Desmodue engine, the Monster 1100 EVO was the most powerful air-cooled Ducati. *Ducati*

GRAND PRIX SUCCESS

CHAPTER 10
A NEW LEASE ON LIFE
PANIGALE, MOTOGP RESURRECTION, THE SCRAMBLER, AND V4 2012–2022

Said to be 90 percent different from the 2011 model, the 2012 Desmosedici GP12 was now 1,000cc, with an aluminum beam frame. *Ducati*

Toward the end of the season, Rossi had a new frame, but it came too late. Following Audi's acquisition of Ducati, the AMG sponsorship decals disappeared. *Ducati*

2012

In April 2012, International Motorcycles agreed to a multi-million Euro acquisition of Ducati by Audi AG, part of the giant Volkswagen Group, the deal was completed in July. Gabriele del Torchio initially continued as CEO, with General Manager Claudio Domenicali also on the board. A record 44,102 motorcycles were sold during 2012, the real growth was in the United States; sales were up 21 percent. The next generation Superbike, the Panigale, was also introduced, along with the 848 Streetfighter, while specific Asian and Australian models included the Monster 795 and 659.

The MotoGP program stalled as Valentino Rossi struggled with the Desmosedici. The last time a new capacity limit was introduced in the premier class was in 2007 and Ducati blitzed it, but on the new 1,000cc D16GP12, Rossi only managed two podium finishes. The power from the V-four Desmosedici was around 255 horsepower at 17,800 rpm, and the frame was an aluminum beam type. Several different frames and swingarms were tried toward the end of the season, but it was to no avail as Rossi had already decided to depart. Nicky Hayden had an even more disappointing season, and in the World Superbike Championship, Checa teamed with Giuliano on the Althea Racing 1198RS12. Although the 1198 was now obsolete, Checa still managed to win four races and finish fourth in the championship points.

1199 Panigale, S, Tricolore, 848 EVO, Corse SE

Replacing the 1198 for 2012 was the most dramatic new model since the 1988 851, the Panigale. Powered by the new Superquadro engine, the Superquadro was completely new other than sharing Fabio Taglioni's trademark 90-degree V-twin layout and desmodromic valve control. The cylinders were rotated backward, with the engine now positioned further forward in the frame, while the horizontally split crankcases allowed the cylinder heads to bolt directly to the crankcase. As on the Desmosedici, the crankshaft ran in plain bearings, and the engine dimensions were extremely oversquare. A new camshaft drive system replaced the venerable belt drive, this featuring a combined chain and gear-drive arrangement, a conventional chain running from the crankshaft to the cylinder head with gears driving the camshafts. With automatic chain tensioning, this system promised improved reliability with reduced maintenance. Another departure from Ducati tradition was the cast-aluminum monocoque frame, although it still incorporated the engine as a stressed member. Completing

RIGHT: The 1199 Panigale S was available in two versions: standard or Tricolore. *Ducati*

BELOW RIGHT: The Superquadro had an innovative chain and gear camshaft drive arrangement. *Ducati*

PANIGALE 1199, S, TRICOLORE, 848 EVO, CORSE SE 2012 Differing from the 1198

Bore (mm)	112
Stroke (mm)	60.8
Displacement (cc)	1,198
Valve actuation	Chain and gear desmodromic double overhead camshafts, four valves per cylinder
Injection	Mitsubishi
Power	195 horsepower at 10,750 rpm
Front suspension	43mm Öhlins (1199S) 50mm Marzocchi (1199)
Rear suspension	Öhlins (1199S) Sachs (1199)
Front brake (mm)	330 dual-disc (848 Corse SE)
Rear tire	200/55x17
Wheelbase (mm)	1,437
Dry weight (kg)	166.5 (Tricolore) 164 (1199, S)

the Panigale's specification were the newest electronic rider aids, Öhlins suspension (on the S), and Brembo brakes. The Panigale was also immediately successful, with 7,500 examples sold during 2012. While the Panigale replaced the 1198, the existing 848 EVO continued unchanged, and it was available this year as a Corse Special Edition (in the same colors as the earlier 1198 Corse).

Diavel, Streetfighter, Multistrada, Hypermotard, Monster

Celebrating Mercedes AMG sponsorship of the Ducati MotoGP team, the Diavel AMG Special Edition was a short-lived model for 2012. Based on the Diavel Carbon, each engine featured individual camshaft timing, with the name of the technician engraved on the left engine casing. The matte black Diavel AMG was discontinued following Audi's purchase of Ducati. Alongside the Diavel and Diavel Carbon this year was the Diavel Cromo, with a gloss-black tank with chrome panels.

The Panigale's monocoque frame and horizontal rear shock absorber were quite radical in design. *Ducati*

Carlin Dunne on his way to winning the 2012 Pikes Peak International Hill Climb. *Ducati*

An addition to the Streetfighter range was the 848, replacing the standard 1098 Streetfighter and powered by an 848cc Testastretta 11° engine. The 1098 Streetfighter S continued unchanged while the best-selling Multistrada lineup now included the 1200S Touring. Carlin Dunne provided the Multistrada its third straight victory at Pikes Peak, dominating the 2012 Pikes Peak International Hill Climb and setting the new course record for motorcycles. The Hypermotard 1100 EVO SP was now available as a Corse Edition, with Corse colors, while the 696, 796, and 1100 EVO Monsters continued unchanged.

STREETFIGHTER 848 2012
Differing from 2011 and the 848 EVO

Power	132 horsepower at 10,000 rpm (848SF)
Front suspension	43mm Marzocchi (848SF)
Rear suspension	Sachs (848SF)
Rear tire	180/60x17 (848SF)
Dry weight (kg)	169 (848SF) 192 (Multistrada)

2013

New models for 2013 included the 1199 Panigale R, Hyperstrada, Hypermotard, and updated Multistrada, sales increasing to 44,287. With 24 percent of sales, the United States was still the dominant market. As a result of better sales, the workforce increased to 1,281, and in April Claudio Domenicali assumed the position of CEO. New management also came to Ducati Corse, ex-BMW Superbike head Bernhard Gobmeier replacing Filippo Preziosi as director. Former 125cc World Champion Andrea Dovizioso joined Hayden on the GP13, and the Desmosedici GP13 was initially similar to 2012, with the same aluminum beam chassis (although several more versions were tried during the year). Again, there was no reward, and without a single podium finish, Ducati had an abysmal year in MotoGP.

In the World Superbike Championship, Team Ducati Alstare fielded factory-supported Panigale 1199RS13s, but

The GP13 was initially similar to 2012's model, but it underwent considerable chassis changes during 2013. *Ducati*

A NEW LEASE ON LIFE | 235

Andrea Dovizioso headed the Ducati 2013 MotoGP team, but his best performance was a fourth place in a wet French Grand Prix. Good results in drier weather were not forthcoming. *Ducati*

PANIGALE 1199R, 848 CORSE SE 2013
Differing from the 1199 and 848 EVO

Dry weight (kg)	165 (1199R)
	167 (848 Corse SE)

with no existing points of reference for the radical monocoque frame and horizontal linkage-operated shock absorber arrangement, setup was problematic. Carlos Checa and Ayrton Badovini struggled for results, Checa finally calling it a day with four rounds to go.

1199 Panigale R, 1199 Panigale, S, Tricolore, 848 EVO, Corse SE

As a homologation model for World Superbike, Ducati released the 1199 Panigale R for 2013. While basically an 1199 Panigale S, the 1199R engine included titanium con rods (increasing the safe rev limit to 12,000 rpm), and the chassis a four-point adjustable swingarm pivot. The R race kit, including a Termignoni racing exhaust system, allowed a 3 percent power increase. The electronic suspension was also racing Öhlins, the front fork a wider World Superbike type to provide optimum brake disc cooling.

The 1199R also included the latest sports-type ABS. The existing 1199 Panigale, S, Tricolore, 848 EVO, and Corse SE continued unchanged.

Hyperstrada and Hypermotard

A new generation Hypermotard was introduced for 2013, the Hypermotard family growing to include the crossover Hyperstrada, this combining motard and touring capability. Powering both was a second-generation Testastretta 11° engine, with ride-by-wire throttle and three maps for specific riding modes. Both the Hypermotard and Hyperstrada shared a

Created as a homologation model for World Superbike, apart from more sophisticated suspension and electronics, the 2013 1199 Panigale R was quite similar to the 1199S. *Ducati*

ABOVE: Combining street and motard capability, the Hyperstrada was a new model for 2013. *Ducati*

BELOW: Sharing the 821 Testastretta engine was the new Hypermotard. *Ducati*

HYPERSTRADA, HYPERMOTARD, SP 2013

Engine	Twin-cylinder four-stroke, liquid-cooled
Bore (mm)	88
Stroke (mm)	67.5
Displacement (cc)	821
Compression ratio	12.8:1
Valve actuation	Belt-driven desmodromic double overhead camshafts, four valves per cylinder
Injection	Magneti Marelli
Power	110 horsepower at 9,250 rpm
Gears	Six-speed
Front suspension	43mm Kayaba 50mm Marzocchi (HMSP)
Rear suspension	Sachs Öhlins (HMSP)
Front brake (mm)	320 dual-disc
Rear brake (mm)	245 disc
Wheels	3.50x17, 5.50x17
Tires	120/70x17, 180/55x17
Wheelbase (mm)	1,490 (HS) 1,500 (HM) 1,505 (HMSP)
Dry weight (kg)	181 (HS) 175 (HM) 171 (HMSP)

new tubular-steel frame, the Hyperstrada including detachable luggage and a centerstand. The Hyperstrada was also available as a low version, with a lower (32.7-inch) seat and shorter suspension. Two versions of the Hypermotard were supplied, the SP with a longer travel Marzocchi fork, Öhlins shock absorber, and Marchesini forged-aluminum wheels.

Multistrada

Evolutionary updates to the Multistrada included a second-generation Testastretta 11° engine, a redesigned front fairing, headlight and screen, and the latest Bosch ABS system. Alongside the higher specification Multistrada S Touring and S Pikes Peak was an S Granturismo, with higher handlebars, screen, and luggage. The S models this year also received Ducati Skyhook Suspension, a semi-active system with a Sachs fork and shock absorber replacing the previous Öhlins.

MULTISTRADA 1200, S 2013
Differing from 2012

Front suspension	48mm Sachs (S)
Rear suspension	Sachs (S)
Dry weight (kg)	194 (S Pikes Peak) 196 (MS) 206 (S Touring) 217 (S Granturismo)

The 2013 Multistrada S Touring included a higher screen and standard luggage. *Ducati*

Diavel, Monster, Streetfighter

The Diavel range now included only the Diavel and Diavel Carbon, and to celebrate twenty years of the Monster, special twentieth anniversary editions of the 696, 796, and 1100 EVO were available. These were standard models, but with different colors for the frame and various brake components. Also available this year was the 1100 EVO Monster Diesel, a collaboration between Ducati and Renzo Rosso, founder of the Diesel brand, its dull green colors emphasizing a military connotation. The two models of Streetfighter were unchanged for 2013.

2014

Four new models were released for 2014, including the 899 Panigale, Limited Edition 1199 Superleggera, third-generation Monster 1200, and an 1199 Panigale S Senna. Sales climbed to 45,117, the United States again leading the way with 8,804 sold. In response to the dismal results in the 2013 MotoGP championship, Ducati lured Luigi Dall'Inga from Aprilia as general manager of Ducati Corse. Ducati started the year with a brand-new bike, opting to stay in the more restrictive Factory category (albeit with special privileges). Engine development saw the power at around 270 horsepower, revving to 18,000 rpm, and during the season the bike became progressively shorter at the front and longer at the rear, with the final update the narrower 14.2 version with a relocated engine.

DIAVEL 2013
Differing from 2012

Dry weight (kg)	205 (Diavel Carbon)

The Monster 1100 EVO twentieth anniversary model included a gold frame and earlier Cagiva-style decals. *Ducati*

Cal Crutchlow replaced Hayden alongside Dovizioso, and the stability generated by Dall'Inga's appointment was immediately evident in the D16GP14's improved performance. Andrea Dovizioso scored Ducati's first pole position in four years at Japan, and consistent results (including a third in Texas) saw him finish fifth overall. Although Crutchlow finished the season strongly (with a podium at Aragon), early crashes eroded his confidence, and he wouldn't complete his two-year contract. After several years in the doldrums, the satellite

Andrea Dovizioso riding to a strong fourth place on the updated GP14.2 at the final race at Valencia. With only one DNF, his season was exemplary. *Ducati*

Pramac Racing Team enjoyed full factory support during 2014, Andrea Iannone's consistent results eventually earning him a factory ride for 2015.

After staring into the abyss during 2013, Ducati Corse resumed control of the World Superbike Team, with new riders Chaz Davies and Davide Giuliano. Under the control of Ernesto Marinelli, the Panigale 1199R enjoyed the lack of restrictors this year, the power climbing to around 217 horsepower at 11,500 rpm. Setup was still tricky, but the Panigale performed much more consistently during 2014. Davies achieved two second places and Giuliano one, finishing sixth and eight respectively in the World Superbike Championship.

BELOW: The Desmosedici GP14 was new—shorter at the front and longer at the rear, with a reshaped fuel tank. *Ducati*

1199 Superleggera, 1199 Panigale, R, Senna, 899

Following in the tradition of the 2008 D16RR, Ducati released the 1199 Superleggera for 2014. Produced in a limited edition of 500, with its cocktail of titanium, magnesium, and carbon-fiber materials, the 200-plus horsepower Superleggera provided the highest claimed power-to-weight ratio of any production motorcycle. Based on the 1199R, the Superquadro engine included for the first time on a Ducati street engine racing Superbike two-ring pistons, the racing kit with titanium Akrapovic exhaust providing a 5 horsepower increase and 2.5-kilogram weight loss. The monocoque frame and Marchesini wheels were magnesium, while the Öhlins suspension and Brembo brakes were racing style. Updated electronics included wheelie control. Dressed in Ducati Corse colors, the 1199 Superleggera continued Ducati's unequaled tradition of building the finest limited-edition race replica motorcycles available.

A NEW LEASE ON LIFE

ABOVE: Ducati Corse returned to World Superbike in 2014, the 1199R benefiting from new regulations removing restrictors. This is Chaz Davies's Panigale. *Ducati*

ABOVE RIGHT: Carrying his hero Kevin Schwantz's No. 34, Davide Giuliano had a mixed 2014 season. *Ducati*

BELOW RIGHT: The 2014 Superbike lineup. The range-topping 1199 Superleggera in Ducati Corse colors heads the 1199R, 1199S, 1199, and 899. *Ducati*

Except for their colors, the 1199 Panigale, S, and R continued unchanged for 2014, the Panigale S Senna joining them. Built as a limited edition of 161 exclusively for the Brazilian market, the Panigale S Senna was finished in the gray with red wheels of the original 1995 916 Senna 1. Also joining the Panigale lineup was the 899 Supermid, with a smaller Superquadro engine. Intended as an entry-level premium performance model, the 899 offered rider-friendly features such as a more comfortable seat, while the suspension included a lighter Showa Big Piston Fork, claimed to offer improved low speed damping. The 899 included the usual Bosch ABS and electronic riding modes, and it was arguably the finest street version in the Panigale range. Ducati also returned to its winning ways in the FIM Superstock 1000, Argentinian rider Leandro Mercado clinching Ducati's fifth crown in this series on an 1199 Panigale R.

1199 SUPERLEGGERA, 899 2014
Differing from the 1199 Panigale R

Bore (mm)	100 (899)
Stroke (mm)	57.2 (899)
Displacement (cc)	898 (899)
Compression ratio	13.2:1 (Superleggera)
Power	200-plus horsepower at 11,500 rpm (Superleggera) 148 horsepower at 10,750 rpm
Front suspension	43mm Showa (899)
Rear wheel	5.50x17 (899)
Rear tire	180/60x17 (899)
Front brake (mm)	320 dual-disc (899)
Wheelbase (mm)	1,426 (899)
Dry weight (kg)	155 (Superleggera) 169 (899)

LEFT: The 1199 Panigale S Senna shared the colors of the first 916 Senna of 1995. *Ducati*

BELOW: The 1199 Superleggera was full of exotic materials, including a magnesium monocoque frame, engine covers, and wheels. *Ducati*

Monster, Diavel, Multistrada, Streetfighter, Hypermotard, Hyperstrada

A new generation liquid-cooled 1200 Monster replaced the air-cooled 1100 EVO for 2014. Derived from the Diavel power cruiser, this continued the style of the earlier S4R and S4RS and was available in two versions: the Monster 1200 and 1200S. Powered by an 1,198cc Testastretta 11° Dual Spark motor, the new chassis included 1199 Panigale-style cylinder head attachments for the Trellis frame. The rear suspension attached directly from the vertical cylinder to the aluminum single-sided swingarm, with a wider rear wheel. The 1200S featured Öhlins suspension and larger diameter front discs, while both the 1200 and 1200S included the Diavel's electronic aids. The Monster 796 and 696 continued unchanged, and a Diavel Dark and Diavel Strada joined the Diavel range, the Strada with a screen, touring seat, and standard luggage. The only Streetfighter available for 2014 was the 848, while the three-model Multistrada and recently introduced Hypermotard and Hyperstrada were unchanged this year.

2015

With several significant new models and model updates, sales increased 22 percent, to 54,800 during 2015. The United States again prevailed in sales numbers (up 14 percent) with the most significant new model the Scrambler. Predominantly built in Thailand, the Scrambler successfully tapped into a new market for modern retro motorcycles and with 16,000 sold was Ducati's best-selling model this year. The 1299 Panigale expanded the parameters for twin-cylinder sportbikes, but as the upper capacity limit for twins in World Superbike remained at 1,200cc, Ducati produced a homologation 1,198cc Panigale R, while other new models included an updated Multistrada (with a Testastretta DVD), limited-edition Diavel Titanium, and Monster 821. Apart from some

Replacing the 848 and completing the Panigale family in 2014 was the 899. *Ducati*

A new generation Monster 1200 replaced the 1100 EVO for 2014. The 1200S had Öhlins suspension. *Ducati*

MONSTER 1200, 1200S 2014
Differing from the Monster 1100

Engine	Twin-cylinder four-stroke, liquid-cooled
Bore (mm)	106
Stroke (mm)	67.9
Displacement (cc)	1,198.4
Compression ratio	12.5:1
Valve actuation	Belt-driven desmodromic double overhead camshafts, four valves per cylinder
Injection	Synerject-Continental with Mikuni throttle bodies
Power	135 horsepower at 8,750 rpm (1200) 145 horsepower at 8,750 rpm (1200S)
Front suspension	43mm Kayaba (1200) 48mm Öhlins (1200S)
Rear suspension	Sachs (1200) Öhlins (1200S)
Front brake (mm)	330 dual-disc (1200S)
Wheels	3.50x17, 6.00x17
Tires	120/70x17, 190/55x17
Wheelbase (mm)	1,511
Dry weight (kg)	182

DIAVEL STRADA 2014
Differing from 2013

Dry weight (kg)	216

specific models for the Asian and Australian market, this year also saw the end of the air-cooled Monster.

As the company hadn't won a MotoGP race since 2010, Ducati was allowed an increased fuel allocation, softer qualifying tires, and more engines during the 2015 season. Under Dall'Inga's direction, the GP15 was redesigned, narrower and shorter, with the engine tilted back, and it was immediately more competitive. Dovizioso and Iannone finished a close second and third at the opening round at Qatar, Dovizioso and Iannone following this with three more second places. Dovizioso's season then went downhill, but Iannone was more consistent, eventually finishing fifth in the championship, with Dovizioso seventh.

No longer burdened by restrictors and required to be closer in specification to the homologated production model, new 2015 World Superbike regulations favored the Desmo twins. The factory Panigale R produced around 8 horsepower less at the beginning of the season, this eventually climbing to 223 horsepower at 11,700 rpm, and at Aragon Chaz Davies provided Ducati its first Superbike win since 2012. He followed this with a double at Laguna Seca and wins in Malaysia and Jerez, ultimately ending second in the championship.

Panigale R

Ducati offered a new limited-edition homologation model for 2015, the Panigale R. Replacing the 1199 Panigale R, as new regulations limited changes to the engine, frame, and electronics, the Panigale R was considerably higher specification. Still displacing 1,098cc, a minimum of 1,000 was required to be built for sale to comply with Superbike regulations. The engine was similar to the 1199 Superleggera, with

LEFT: Andrea Iannone was more consistent than Dovizioso during 2015, the GP15 notable for the small winglets to provide downforce and prevent wheelies. *Ducati*

BELOW: Chaz Davies broke Ducati's race-winning drought in World Superbike, taking five victories during 2015. *Ducati*

The Desmosedici was redesigned for 2015, now smaller and more tightly packaged. It was still the only MotoGP contender to use a carbon-fiber swingarm. *Ducati*

BELOW: New World Superbike regulations for 2015 required the factory racing bikes to be more similar to the homologation version. The production Panigale R is in front of the factory bikes of Giuliano (left) and Davies (right). *Ducati*

A NEW LEASE ON LIFE | 243

PANIGALE R 2015
Differing from the 1199 Panigale R 2014

Compression ratio	13.2:1
Power	205 horsepower at 11,500 rpm
Wheelbase (mm)	1,442
Dry weight (kg)	162

1299 PANIGALE, S 2015
Differing from the 1199 Panigale 2014

Bore (mm)	116
Displacement (cc)	1,285
Compression ratio	12.6:1
Power	205 horsepower at 10,500 rpm

two-ring pistons, titanium valves, and lightweight crankshaft. New electronics included a Bosch inertial measuring unit that provided corning ABS. The chassis also featured an adjustable swingarm pivot and Öhlins mechanical suspension.

1299 Panigale, S

After only three years, the 1199 Panigale evolved into the 1299 Panigale for 2015. The Superquadro engine featured the largest pistons ever for a Supersports road bike, and as the 1299 produced 10 percent more torque, it was more responsive and easier to ride than the 1199. New electronic innovations also improved the rideability, while a slightly steeper steering rake, less trail, and a lower swingarm pivot increased agility and rear traction. Styling updates extended to a wider fairing, with wider front air intakes, a more protective windshield, and a new tail unit. The twin headlamps were integrated as part of the front intake ports. As on the Panigale R, the new electronic kit included the inertial measurement unit, while the 1299 Panigale S featured semi-active Öhlins suspension.

Multistrada 1200, 1200S, D-Air

Apart from a redesigned fairing and narrower seat, styling-wise the new Multistrada was similar to before, but technically it was all new. The Testastretta included desmodromic variable timing (DVT), the first time variable valve timing was applied to the inlet and exhaust camshafts of a production motorcycle engine. Electronic updates included the new Bosch inertia measurement unit that measured roll and pitch angles in three axes and electronic cruise control. Other innovations this year saw the introduction of D-Air on the 1200S, an integrated air bag system developed in conjunction with Dainese. A new frame provided improved ground clearance, and four different versions were available: Touring, Sport, Urban, and Enduro, each offering a range of specific equipment.

The 1299 Panigale not only featured a larger displacement engine, but also included a new electronics package and a restyled front fairing. *Ducati*

Monster

With more than 290,000 Monsters sold since 1993, the Monster range was still pivotal for Ducati and this year the liquid-cooled 821 replaced the air-cooled 796 and more exclusive Monster Stripe 1200S and 821 versions were introduced. These included a small fairing and double white stripes on the front fender, tank, and seat cover. The Monster 1200 and 1200S continued unchanged.

LEFT: The Multistrada also received a mild facelift for 2015 and incorporated a number of electronic updates. *Ducati*

ABOVE: The Multistrada's Testastretta engine included hydraulically operated desmodromic variable valve timing. *Ducati*

Powered by the Hypermotard's second-generation Testastretta 11° engine, the Monster 821 followed the style of the 1200, with a trellis frame attached to the cylinder heads. While the Kayaba front fork and Sachs shock were unchanged, the 30mm shorter swingarm was now double sided. The compact Monster 821 emphasized ergonomics, with a higher and closer handlebar and low seat. An accessory 745mm seat was Ducati's lowest ever, and a black Monster 821 Dark was also available.

Scrambler

The most anticipated new release for 2015 was the Scrambler, a contemporary take on the popular Scrambler that ran from 1962 to 1975. Unlike some other retro models, the Scrambler was very much a modern motorcycle, much as it would have been if production had continued. Four versions were

MULTISTRADA 1200, 1200S, D-AIR 2015
Differing from 2014

Compression ratio	12.5:1
Injection	Bosch
Power	160 horsepower at 9,500 rpm
Front suspension	48mm fork
Wheelbase (mm)	1,529
Dry weight (kg)	209 (MS) 212 (S) 213 (D-Air)

MONSTER 821 2015
Differing from the Monster 1200 and Hypermotard 2014

Injection	Continental with Mikuni throttle bodies
Power	112 horsepower at 9,250 rpm
Front brake (mm)	320 dual-disc
Rear wheel	5.50x17
Rear tire	180/60x17
Wheelbase (mm)	1,480
Dry weight (kg)	179.5

The new middleweight Monster for 2015 was the 821. The engine was now a liquid-cooled Testastretta and the swingarm double sided.

The basic Scrambler was the Icon, the bright yellow colors harking back to the Scramblers of the 1970s. *Ducati*

SCRAMBLER 2015

Engine	Twin-cylinder four-stroke, air-cooled
Bore (mm)	88
Stroke (mm)	66
Displacement (cc)	803
Compression ratio	11:1
Valve actuation	Belt-driven desmodromic single overhead camshaft, two valves per cylinder
Power	75 horsepower at 8.250 rpm
Gears	Six-speed
Front suspension	41mm Kayaba
Rear suspension	Kayaba
Front brake (mm)	330 single-disc
Rear brake (mm)	245
Wheels	3.00x18, 5.50x17
Tires	110/80x18, 180/55x17
Wheelbase (mm)	1,445
Dry weight (kg)	170

available: the standard Icon, Urban Enduro, Full Throttle, and Classic.

Derived from the 796 Monster, the air-cooled twin featured camshafts with an 11-degree valve overlap, while the frame was the usual steel trellis with an aluminum swingarm. Modern components included an inverted front fork, front radial-mount brake caliper with Bosch ABS, and monoshock rear suspension. Both the Urban Enduro and Classic featured wire-spoked wheels. Retro features included a "born in 1962" inscription on the fuel filler cap and an ignition key that inserted into the headlamp as on the original, and a classic round headlamp. Whereas many retro bikes struggled to match classic style with modern function, with the Scrambler Ducati found the right formula. Light and compact, with excellent detailing, the Scrambler was a success.

Diavel, Hypermotard, Streetfighter

While the Diavel and Diavel Carbon continued much as before, an exclusive limited-edition Titanium was also available this year. This included a number of distinctive cosmetic features, from titanium headlight and tank covers to a number of carbon-fiber components. Limited to 500 bikes, each came with a unique numbered identification plaque. The Hypermotard, Hypermotard S, Hyperstrada, and Streetfighter 848 were unchanged.

2016

As Ducati prepared for its seventieth anniversary in 2016, it released nine new models. Early announcements included the Monster 1200R and Diavel Carbon with seven more models debuting at EICMA in November 2015: the XDiavel, two new Scramblers, Hypermotard and Hyperstrada 929, Panigale 959, and Multistrada 1200 Enduro.

The Diavel Titanium was a limited edition of 500 units. In addition to wider carbon-fiber air intakes, it had a hand-stitched Alcantara seat with leather inserts. *Ducati*

Sales in 2106 increased by 1.2 percent to 55,451 motorcycles. The United States remained the largest market with 8,787 units, with Italy second, ahead of Germany.

In MotoGP, the new Desmosedici GP16 employed a more compact V-4 engine producing some 280 horsepower at 18,000 rpm. It was consistently the fastest bike on the track. At the Austrian round, Iannone provided Ducati its first MotoGP victory since 2010.

Davies and Giuliano headed the factory Aruba.it Racing World Superbike Championship team riding the Panigale 1199R.

LEFT: The GP16 received a new chassis and fairing with more sophisticated wings. The wheels were 17-inch this year and the tires Michelin. Now in his fourth and final year with Ducati, Andrea Iannone won the 2016 Austrian Grand Prix but crashed heavily at Silverstone two rounds later, ruining his Championship hopes. *Ducati*

ABOVE: Chaz Davies won eleven World Superbike races on the Panigale 1199R during 2016 and finished third overall in the World Championship. Here he is on his way to victory at Laguna Seca. *Ducati*

BELOW: Taking the cruiser concept to a new level was the larger-displacement XDiavel S. As the water pump was moved from the left, the pipes were no longer visible and the engine looked cleaner. *Ducati*

XDiavel, XDiavel S, and Diavel Carbon

Evolution of the Diavel family continued with a new Diavel Carbon. It featured a redesigned exhaust, wheels, and seat, but the XDiavel was even more intriguing. Endeavoring to merge two separate forms—the pure cruiser and traditional sports style—the XDiavel included cruiser-style forward foot pegs for low-speed maneuverability in a high horsepower package. Three numbers summed up the XDiavel; 5,000, 60, and 40. 5,000 was the rpm at which the Testastretta DVT engine achieved its maximum torque of 95 lb-ft; 60 the total of different rider ergonomic configurations; and 40 the degrees of maximum attainable lean angle. The Testastretta DVT engine capacity was increased to 1262cc, with the water pump relocation inside the cylinder V improving the left side aesthetics. Individual personalization included four different footrest positions, five alternative seats, and three handlebars. The frame, suspension, brakes, and the chassis setup were designed to combine sports riding without compromising highway comfort. Completing the specification was a cruiser-style belt final drive. Also available as an S version in gloss rather than matte black, the XDiavel included Brembo M50 front brake calipers, a special seat, and aluminum engine belt covers.

Monster 1200R

With the advent of the higher horsepower naked BMW S1000R and KTM 1290 Super Duke R, Ducati upped the ante with the Monster 1200R for 2016. A power increase came courtesy of a higher compression ratio, larger diameter oval throttle body, and new exhaust, delivered through a slipper clutch,

XDIAVEL, S 2016 Differing from the Diavel

Stroke (mm)	71
Displacement (cc)	1262
Compression ratio	13:1
Injection	Bosch
Power	156 horsepower at 9,500 rpm
Wheelbase (mm)	1,615
Dry weight (kg)	220

MONSTER 1200R 2016
Differing from the Monster 1200S

Compression ratio	13:1
Power	160 horsepower at 9,250 rpm
Rear tire	200/55x17
Wheelbase (mm)	1,509
Dry weight (kg)	180

The first of nine new models released for 2016 was the high performance Monster 1200R. *Ducati*

eight wheel spin tailoring profiles, and three riding modes. The Öhlins suspension was raised at both ends to improve cornering clearance, the rear tire was wider, and a small fairing added, along with a new tailpiece.

Multistrada 1200 Enduro, Pikes Peak

A new Multistrada 1200 Enduro joined the unchanged Multistrada, S, and D-Air. With spoked 19- and 17-inch wheels, semi-active electronic Sachs suspension with more travel, and a 30-liter fuel tank, the 1200 Enduro was designed for the maxi road enduro segment and provided a range exceeding 280 miles. The Pikes Peak special edition also made a return and was based on the 1200S with Öhlins suspension and racing-inspired three spoke wheels.

Panigale 959

Just as the 1299 replaced the 1199 (this time after only *two* years), the Panigale 899 made way for a larger 959. Continuing a tradition where title didn't reflect the actual displacement, the 955cc engine included a longer stroke and made slightly more power, while other updates included the 1299's wider nose fairing and screen, larger section front air intakes, and split tailpiece.

Hypermotard 939, Hypermotard 939SP, Hyperstrada 939

Another replacement for a recent model was the introduction of the 939 Hypermotard range, replacing the 821 introduced three years earlier. The entire range shared the new Testastretta 11° engine, again the displacement of 937cc not represented in the model designation. Engine updates included a bore increase, a 42mm diameter crankpin, and a slightly higher compression ratio. An oil cooler was also added to the standard water radiator. As before, the Hypermotard 939SP included Öhlins suspension while the Hyperstrada 939 came equipped with 50-liter removable bags and a center stand.

Ducati Scrambler Flat Track Pro and Sixty2

Joining the highly successful existing four-model Scrambler range were the Flat Track Pro and Sixty2. Based on the Scrambler Full Throttle and inspired by the American AMA Pro Flat Track 2015 Championship bikes of Troy Bayliss and

MULTISTRADA 1200 ENDURO 2016
Differing from the Multistrada 1200

Front brake (mm)	320 dual disc
Wheels	3.00x19, 4.50x17 (Spoked)
Tires	120/70x19, 170/60x17
Wheelbase (mm)	1,556
Dry weight (kg)	225

Joining the Multistrada lineup for 2016 was 1200 Enduro with a larger fuel tank and wire-spoked wheels. *Ducati*

959 PANIGALE 2016
Differing from the 899

Stroke (mm)	60.8
Displacement (cc)	955
Power	157 horsepower at 10,500 rpm
Wheelbase (mm)	1,431
Dry weight (kg)	176

Separating the 959 Panigale from the outgoing 899 was a new exhaust system and restyled front fairing. *Ducati*

The Hypermotard 939SP had MotoGP-inspired colors and top-of-the-line Öhlins suspension. *Ducati*

HYPERSTRADA, HYPERMOTARD, SP 2016
Differing from 2015

Bore (mm)	94
Displacement (cc)	937
Compression ratio	13.1:1
Power	113 horsepower at 9,000 rpm
Wheelbase (mm)	1,485 (HS) 1,493 (HM) 1,498 (HMSP)
Dry weight (kg)	187 (HS) 181 (HM) 178 (HMSP)

Johnny Lewis, the Flat Track Pro featured a new racing yellow gas tank, a side-mounted plate holder, and unique nose fairing and short front fender. Standard equipment included accessorized footrests, sprocket cover, rearview mirrors, and front brake fluid filler cap. The Sixty2, named after the year of the original Scrambler, was aimed at an entry-level clientele, with a smaller capacity engine, smaller rear wheel and tire, steel swingarm, and dedicated colors and graphics.

SCRAMBLER SIXTY2 2016
Differing from the Scrambler

Bore (mm)	72
Stroke (mm)	49
Displacement (cc)	399
Compression ratio	10.7:1
Power	41 horsepower at 8,750 rpm
Front suspension	41mm Showa
Front brake (mm)	320 single disc
Rear Wheel	4.50x17
Rear Tire	160/60x17
Wheelbase (mm)	1,460
Dry weight (kg)	167

A new Scrambler for 2016 was the Flat Track Pro, based on the Full Throttle and inspired by Troy Bayliss' American AMA Pro Flat Track 2015 Championship bike. *Ducati*

During the 2017 year the GP17 received this new hammerhead-style fairing. But while Andrea Dovizioso won with it in Austria and Japan, he generally still used the earlier type. *Ducati*

2017

Eight new models were released for 2017. Following engine rationalization, the Testastretta now produced in 1,262, 1,198, and 937cc displacements, with the Desmodue available only as 803 and 399cc. The Superquadro continued in 1,285, 1,198, and 955cc displacements. The only surviving 821 Testastretta was the Monster 821, and the Hyperstrada was discontinued. Motorcycle sales totaled 55,871 units during 2017, with the Scrambler the most popular at 14,061 deliveries. The US again led the way with 8,898 units, but the Italian market was close behind, climbing 12 percent to 8,806 units.

Multiple MotoGP World Champion Jorge Lorenzo signed a two-year deal with Ducati for 2017 and 2018. The Desmosedici GP17 was a development of the GP16. The 2017 fairing was initially similar to the previous year's (minus external winglets), but this changed during the season with the appearance of a controversial new hammerhead fairing. With these developments, the 2017 season was Ducati's most successful in a decade: Dovizioso won six races and finished second in the World Championship.

Former MotoGP rider Marco Melandri joined Chaz Davies in the World Superbike Aruba.it Racing team. The 1,198cc Panigale R Superbike specifications changed little from 2016. With more than 215 horsepower at 12,000 rpm, the factory Panigale R provided strong competition but the two riders weren't consistent enough to win the Championship.

Although it was little changed from 2016, the Panigale 1199R was even more competitive in 2017. Chaz Davies used it to post seven race wins and finish second in the 2017 World Superbike Championship. *Ducati*

Panigale R rider Glenn Irwin provided Ducati its first-ever win in the North West 200. Veteran British Superbike rider Shayne Byrne also pulled off a surprising victory in the British Superbike Championship. The Panigale R was also victorious in the FIM Superstock 1000 Championship, with Michael Ruben Rinaldi securing the title.

1299 Panigale S Anniversario

Unveiled by Casey Stoner at the 2016 World Ducati Week, the 1299 Panigale S Anniversario was an early release for 2017, created to celebrate the company's ninetieth anniversary in a

A NEW LEASE ON LIFE | 251

1299 PANIGALE S ANNIVERSARIO
Differing from the 1299 S Panigale

Weight	164 kg (362 lb)

Inspiration for the 1299 Panigale S Anniversario's white, black, and red bodywork and gold wheels came from Ducati's racing bikes. *Ducati*

1299 SUPERLEGGERA
Differing from the 1299 Panigale S

Compression ratio	13.0:1
Power	215 horsepower at 11,000 rpm
Front suspension	43mm Öhlins FL936 upside-down fork
Rear suspension	Öhlins TTX36 shock absorber
Wheelbase (mm)	1,456
Dry weight	156 kg (344 lb)
Dry weight (kg)	167

The second series of Superleggera was the 1299, which featured a carbon-fiber chassis. *Ducati*

limited edition of only five hundred. The top triple clamp and steering head insertswere machined from solid aluminum alloy, and the top triple clamp laser-etched with the bike's production number. The steering head inserts shifted the front wheel forward 5mm, providing the same geometry as the Panigale R, and carbon-fiber components reduced the overall weight by 2.5 kilograms. The electronics included all-new traction and wheelie control. Each bike came with a racing kit with a titanium Akrapovič Racing exhaust system.

1299 Superleggera

The 1299 Superleggera was the second in the Superleggera series, again in a limited edition of five hundred. The most powerful production Ducati yet produced, it was also the first production bike to feature a carbon fiber frame, swingarm, subframe, and wheels. All of this combined to make it 40 percent lighter than the 1299 Panigale. Completing the advanced chassis specification were racing-quality Öhlins suspension and Brembo brakes. Updates to the 1,285cc Superquadro engine included sand-cast crankcases, titanium conrods, larger titanium valves, and a titanium Akrapovič exhaust, similar to the official factory WSBK Panigale.

A new electronics package included a six-axis Inertial Measurement Unit (6D IMU) with Ducati Slide Control (DSC) and Ducati Power Launch (DPL). Each example came with a track kit.

SuperSport

The SuperSport resurrected another great name from Ducati's past, looking back to the original Super Sport (1973–1982) and Supersport (1989–2006). The SuperSport was less

SUPERSPORT, SUPERSPORT S
Differing from the Hypermotard

Compression ratio	12.6:1
Front suspension	43mm Marzocchi; 48mm Öhlins (S)
Rear suspension	Sachs, Öhlins (S)
Wheelbase (mm)	1,478
Dry weight (kg)	184

hard edged than more recent superbikes and designed to offer sports-touring capability. A redesigned fairing included a height-adjustable Plexiglas screen that could be set to two different positions over 50mm of travel.

Powering the SuperSport was an adapted Hypermotard twin-cylinder 937cc Testastretta 11° engine. The power was unchanged and, to meet Euro 4 emission standards, the 2-1-2 54mm exhaust system included a large belly silencer with two stacked end-cans. The crankcase and cylinder heads were redesigned externally, and the clutch was an oil-bath slipper type. Three riding modes (Sport, Touring, and Urban) were offered.

The steel trellis frame was based on the Monster 821, with the engine a fully stressed chassis component. This connected to the front cylinder head, the rear steel subframe to the rear cylinder head, and the single-sided die-cast aluminum swingarm to the crankcase. Two versions of the SuperSport were available, the standard version and the higher-specification SuperSport S. Blending real-world sportbike performance with a more comfortable riding position and softer suspension, the new SuperSport created a new market segment for Ducati.

Higher handlebars and a lower seat helped the SuperSport provide all-around sporting ability. *Ducati*

Multistrada 950 and 1200 Enduro Pro

With fifty thousand units sold since its introduction in 2010, the Multistrada 1200 was one of Ducati's best-selling models. This success begged a midsize version. Already adapted for the SuperSport, the new Hypermotard Testastretta 11° engine proved an ideal powerplant for the downsized Multistrada. The Euro 4-rated engine received a redesigned oil system and new desmodromic cylinder heads with revised porting and secondary air intake ducts. The 950 frame was a large-diameter thin tube trellis with two lateral aluminum subframes. The double-sided aluminum swingarm and wheel sizes were similar to those of the 1200 Enduro.

Taking styling cues from the two larger Multistradas, the 950 fairing featured an adjustable screen while the narrow seat and high-exit silencer was comparable to the 1200 Enduro. Four optional packages were available: Touring, Sport, Urban, and Enduro. Intended as a budget version, the

The Multistrada 950 could be ordered with optional equipment packs. This bike has the Touring and Urban packages. *Ducati*

A NEW LEASE ON LIFE | 253

MULTISTRADA 950, 1200 ENDURO PRO
Differing from the Multistrada 1200 Enduro

Bore (mm)	94 (950)
Stroke (mm)	67.5 (950)
Displacement (cc)	937 (950)
Compression ratio	12.6:1 (950)
Injection	Bosch electronic with 53mm throttle bodies (950)
Power	113 horsepower at 9,000 rpm (950) 152 horsepower at 9,500 rpm (1200 Pro)
Front suspension	48mm Kayaba (950)
Wheels	3.00x19, 4.50x17 (950 and 1200 Pro)
Tires	120/70x19, 170/60x17 (950 and 1200 Pro)
Wheelbase (mm)	1,594 (950 and 1200 Pro)
Dry weight (kg)	204 (950) 232 (1200 Pro)

MONSTER 1200, 1200S 2017
Differing from the Monster 1200 (2016)

Compression ratio	13:1
Power	150 horsepower at 9,250 rpm
Wheelbase (mm)	1,485
Dry weight (kg)	187 (1200) 185 (1200S)

The new Monster 1200S employed an exposed frame to recreate the minimalist look of the original 1993 Monster. *Ducati*

Multistrada 950 still offered acceptable performance without its big brother's newer electronic aids.

The Multistrada 1200 Enduro Pro joined the Multistrada lineup for 2017. More focused on off-road riding than the 1200 Enduro, the Pro's specific features included a rough-finished, sand-colored front end and tank cover, two-tone seat, and a black subframe and clutch/alternator covers. Other Pro features included Touratech bull bars with LED lights, a low screen, and a Termignoni titanium exhaust.

Monster 1200, 1200S

More powerful, updated, and restyled, the Monster 1200 and 1200S were introduced this year. The elemental style of Galluzzi's original 1993 version served as inspiration for the new 1200, which featured a sleeker fuel tank and redesigned tail. Attached to the engine was a new single-sided swingarm and die-cast aluminum footpegs. Both versions included a comprehensive electronics package with three riding modes (Sport, Touring, and Urban), an IMU for Bosch Cornering ABS, and Ducati Wheelie Control (DWC) systems. Also new were the dual parabola headlight, an instrument panel now attached to the handlebar risers, and new switches. Part of the redesign reduced the number of plastic parts to recreate the original Monster's minimalist look.

Monster 797

For 2017, Ducati reintroduced an entry-level air-cooled Monster. Sharing the Scrambler's 803cc Desmodue engine, the 797 included redesigned cooling fins reminiscent of the first-generation Monster 900. The frame was the traditional tubular trellis with a die-cast aluminum twin-sided swingarm and laterally mounted shock absorber. The styling intentionally recalled the original 1993 Monster. A 797+ version included a flyscreen and passenger seat cover. Optional accessories for both models included a Sport and Urban pack.

Diavel Diesel

Following on from the 2013 1100 EVO Monster Diesel, Ducati again teamed with Diesel to create the limited edition Diavel Diesel. Specific features included a hand-brushed stainless steel tank cover with visible welding and rivets, a leather

The Scrambler Café Racer also echoed an earlier era and carried Italian racer Bruno Spaggiari's #54. *Ducati*

seat, red Brembo front brake calipers, and a vintage style LCD dashboard. With a sly wink, the numbered edition was limited to 666 units.

Scrambler Desert Sled and Café Racer

Two additional Scrambler variations appeared in 2017: the Desert Sled and Café Racer. Stealing a term associated with 1960s West Coast Triumph off-road race bikes, the Desert Sled was a dirt-style Scrambler, reminiscent of the 1970s 450 R/T. The exhaust system remained low slung, but an engine sump guard protected the header pipe. The frame was reinforced around the swingarm pivot and steel plates on either side incorporated the rear engine mount, swingarm pivot points, and lower footpeg mounts. The new reinforced aluminum swingarm was longer and straighter than what was found on the standard Scrambler and the aluminum shock absorber included a separate gas reservoir. The wire-spoked wheels included gold anodized rims and were fitted with crossover rally-style tires. Other off-road features included a headlight mesh guard and high fenders. To provide acceptable ground clearance and suspension travel, the overall weight and dimensions were increased considerably over the standard Scrambler.

Joining the Desert Sled was the Scrambler Café Racer. With its low handlebars and rear-set footpegs, this was Ducati's interpretation of the popular street racer. The café racer was initially a street racer style that originated in England during the 1960s. Ducati was among the first manufacturers to offer a factory version with its 250 Mach 1. Over the years Ducati has offered many similar café-style machines.

Arguably more stylish than functional, the Scrambler Café Racer was based on the Scrambler Icon with its black-finished engine and machined cooling fins. The Café Racer added a dual Termignoni exhaust with black anodized aluminum cover. Paint was a traditional British black and gold complemented by a small fairing. The number plate holders on each side carried the number 54 in honor of factory rider Bruno Spaggiari, who had raced with this number on the factory 250 Desmo in the 1968 Mototemporada Romagnola road races in Italy.

MONSTER 797
Differing from the Scrambler

Front suspension	43mm Kayaba
Rear suspension	Sachs
Front brake (mm)	320 twin disc
Front Wheel	3.50x17
Front Tire	120/70x17
Wheelbase (mm)	1,435
Dry weight (kg)	175

SCRAMBLER DESERT SLED AND CAFÉ RACER
Differing from the Scrambler

Front suspension	46mm Kayaba (Desert Sled)
Wheels	3.00x19 and 4.50x17 (Desert Sled) 3.50x17 (Café Racer *front*)
Tires	120/70x19 and 170/60x17 (Desert Sled) 120/70x17 (Café Racer *front*)
Wheelbase (mm)	1,505 (Desert Sled) 1,436 (Café Racer)
Dry weight (kg)	191 (Desert Sled) 172 (Café Racer)

The Scrambler Desert Sled was inspired by the Californian off-road "desert sled" racers of the 1960s. *Ducati*

A NEW LEASE ON LIFE | 255

ABOVE: This GP18 Desmosedici is Dovizioso's bike. He began the season with the aero fairing and a win at Qatar. *Ducati*

BELOW: The final version of the two-cylinder Panigale was this 1299 Panigale R Limited Edition. *Ducati*

1299 PANIGALE R FINAL EDITION
Differing from the 1299 Panigale and Panigale R

Compression ratio	13:1
Power	209 horsepower at 11,000 rpm
Wheelbase (mm)	1,443
Dry Weight (kg)	168

2018

Many new models appeared in Ducati's 2018 lineup, while this year saw the end of a forty-seven-year tradition: factory racing 90-degree V-twins. Staying on in World Superbike for one more year, the factory Panigale R would soon give way to its four-cylinder replacement, the Panigale V4, which was introduced this year. Ducati released the 1299 Panigale R Final Edition to celebrate the end of the V-twin's reign. Despite a small decline in worldwide sales of motorcycles over 500cc, the company managed to maintain sales above fifty thousand units for the fourth consecutive year. In this period the new Panigale V4 was the world's best-selling Superbike.

Dovizioso and Lorenzo raced in the Ducati MotoGP team. Coming out of the successful 2017 season, the GP18 represented a further evolution, its engine now more powerful while also refined to improve rideability. Updates included an Öhlins TSB 46 CF "Blackfisken" 46mm front fork with carbon outer upper tubes and revised aerodynamics. Thanks to his excellent results during 2017, Petrucci had earned a ride on the OCTO Pramac team GP18, with Jack Miller replacing Scott Redding on the GP17. Ducati won seven GPs, its highest win tally in a single season since Casey Stoner's glory year a decade earlier. Dovizioso finished second overall and Lorenzo ninth, one place behind Petrucci.

In the World Superbike Championship, Marco Melandri continued alongside Chaz Davies on the Aruba.it Racing Panigale R. The Panigale R raced for the last time in World Superbike as new technical regulations mandated a maximum rpm of 12,400. Davies managed two race wins to finish second overall in the championship. Melandri had a disappointing season and finished fifth overall.

1299 Panigale R Final Edition

The 1299 Panigale R Final Edition was released as a commemorative model at the moment when Ducati's tradition of factory racing twins was coming to an end. This bike was available as a numbered (but not limited) series, a Euro 4-compliant offering based on the 1299 Superleggera and Panigale R. The Final Edition Superquadro engine featured a lightened crankshaft with a larger crankpin and titanium con rods. Along with larger valves, the revised intake and exhaust ducts were combined with new higher lift camshafts. The titanium Akrapovič exhaust with a racing-style dual muffler was also new. The electronics package included the Bosch IMU and ABS Cornering, DWC EVO, Ducati Traction Control EVO (DTC EVO), and Engine Brake Control (EBC). The chassis was a die-cast aluminum monocoque frame and Öhlins mechanical suspension similar to the Panigale R. The 1299 Panigale R Final Edition was historically significant as the final iteration of Ducati's premier V-twin sporting model.

Panigale V4, V4S, V4 Speciale

To compensate for the twin-cylinder Panigale's disappointing turn in World Superbike racing when compared with earlier twins, the Panigale V4 appeared in 2018 as its replacement. Developed through fifteen years of MotoGP competition, this new 90-degree V-4 was designed to power future range-topping sport models. Three versions were offered: the Panigale V4, Panigale V4 S with higher specification suspension, and the limited-edition monoposto Panigale V4 Speciale. The compact Desmosedici Stradale V4 engine shared the cylinder heads, dimensions and frame geometry with the Desmosedici GP. The engine was rotated rearward by 42 degrees, optimizing weight distribution, allowing for larger radiators, and bringing the swingarm pivot as far forward as possible.

Similar to the MotoGP Desmosedici engines, the counterrotating three main-bearing crankshaft was offset at 70 degrees, allowing for a "Twin Pulse" ignition sequence. The camshaft drive, however, was similar to the two-cylinder Superquadro, employing a combination of chain and gears. The cam covers, oil sump, alternator cover, and two-piece clutch cover were die-cast magnesium.

The "Front Frame" chassis was an evolution of the previous Panigale's monocoque and attached directly to the front and rear cylinder heads, with the engine also acting as a fixing point for the rear suspension and a fulcrum point for the swingarm. The shock absorber was positioned more conventionally and front braking was effected by new Brembo Stylema Monobloc calipers.

The Panigale V4 was the most advanced production Ducati yet. As part of its promotion, Panigale V4Ss were provided for a "Race of Champions" at Misano at World Ducati Week

PANIGALE V4, V4S, V4 SPECIALE
Differing from the Multistrada 1200 Enduro

Engine	Four-cylinder, 4-stroke, liquid-cooled
Bore (mm)	81
Stroke (mm)	53.5
Displacement (cc)	1103
Compression ratio	14:1
Valve actuation	Desmodromic double overhead camshafts, 4 valves per cylinder
Injection	Electronic 52mm throttle bodies
Power	214 horsepower at 13,000 rpm
Gears	Six-speed
Front suspension	43mm Showa BPF 43mm Öhlins NIX30 (S, Speciale)
Rear suspension	Sachs Öhlins TTX36 (S, Speciale)
Front brake (mm)	330 dual disc
Rear brake (mm)	245 disc
Wheels	3.50x17, 6.00x17
Tires	120/70x17, 200/60x17
Wheelbase (mm)	1,469
Dry weight (kg)	175 174 (S, Speciale)

With its single-sided swingarm the Panigale V4 continued the style of earlier Ducati Superbikes. Its V4 Stradale engine was based on the Desmosedici MotoGP V4. This is the higher specification V4S with Öhlins suspension. *Ducati*

in July 2018. Michele Pirro won the race from an illustrious field and the thirteen Panigale V4Ss were subsequently auctioned on eBay.

959 Panigale Corse

As there was no longer a higher-specification Panigale in the lineup following the demise of the 1299 Panigale S, Ducati introduced the 959 Panigale Corse for 2018. The Euro 4, approved for Europe, and the non-Euro 4 for the US were the two versions produced. The Euro 4 version featured a twin titanium muffler by Akrapovič, while US versions retained the under-engine muffler. The aluminum double-sided swingarm was carried over from the 959 but with higher-spec Öhlins suspension.

Multistrada 1260, 1260 S, 1260 S D-Air, 1260 Pikes Peak

A number of updates for the Multistrada came in 2018: the engine's new Ducati Testastretta Desmodromic Variable Timing (DVT), new chassis, more advanced electronics, and new styling. In addition, this bike featured new fairing panels and lighter wheels. The engine, similar to the XDiavel's, had a longer stroke that required new connecting rods, crankshaft, and cylinders. The DVT system's recalibration maximized low- to-mid-range torque. This meant there was a slight reduction in maximum power, but torque increased by 18 percent with 85 percent available from 3,500 rpm.

There were further updates that distinguished the Multistrada for this year, such as a restyled timing belt and outer engine covers, a tubular steel trellis frame, and longer single-sided die-cast aluminum swingarm. The frame's two lateral subframes were closed off by a rear load-bearing element made of fiberglass-reinforced plastic, a design that offered maximum torsional rigidity. While the Kayaba and Sachs suspension was unchanged, the lighter wheels were new. As an electronic aid, the Multistrada featured a Vehicle Hold Control (VHC) system, which applied rear-wheel braking for hills.

The Multistrada 1260 Pikes Peak returned after four years, this time with higher specifications. It featured racing-inspired colors, a red frame, black forged aluminum wheels, and mechanical Öhlins suspension.

The 959 Panigale Corsa was offered in two variants. The Euro-spec version here had twin Akrapovič mufflers. *Ducati*

959 PANIGALE CORSE
Differing from the 959 Panigal

Power	150 horsepower at 10,500 rpm
Front suspension	43mm Öhlins NIX30
Rear suspension	Öhlins TTX36
Dry weight (kg)	175.5

MULTISTRADA 1260, S, D-AIR, PIKES PEAK
Differing from the Multistrada 1200

Stroke (mm)	71.5
Displacement (cc)	1262
Compression ratio	13:1
Power	158 horsepower at 9,500 rpm
Front suspension	48mm Öhlins (Pikes Peak)
Wheelbase (mm)	1,585
Dry weight (kg)	206 (Pikes Peak)

LEFT: Although it looked similar to the Multistrada 1200, The Multistrada 1260 was significantly redesigned. *Ducati*

ABOVE: The Scrambler 1100 Sport included Öhlins suspension and custom wheels, handlebar, and seat. *Ducati*

Scrambler 1100, 1100 Special, 1100 Sport, Street Classic, and Mach 2.0

An 1100 filled out the Scrambler range for 2018, this time powered by an updated version of 2013's Monster 1100 engine. With its twin spark plugs per cylinder, the new 1100 offered smoother power delivery and minimal emissions.

This bike's transmission, engine, and frame had undergone important updates. The clutch was a hydraulically actuated wet multiplate type with a servo-assisted slipper function, while the engine included machine-finished aluminum clutch, alternator, and belt covers. The Scrambler 1100 was a larger overall bike compared with the 800, thanks to its new twin upper spar steel trellis frame with an aluminum rear subframe.

Three versions of the 1100 were available at first: the basic Scrambler 1100 in "62 Yellow" or Shining Black; the classic Scrambler 1100 Special with black-spoked wheels, gold anodized front fork sleeves, chrome exhausts, and aluminum mudguards; and the custom Scrambler 1100 Sport.

The Scrambler 1100 was the first in the Scrambler family to feature DTC. Three riding modes were also offered: Active (full power), Journey (for everyday use), and City (75 horsepower).

A Street Classic and Mach 2.0 joined the standard Scrambler range for 2018. The Street Classic was similar to the Classic, with wire-spoke wheels and a brown seat, while the Mach 2.0 was a product of designer Roland Sands. Inspired by the 250 Mach 1 of 1964–1966, the Mach 2.0's graphics and colors evoked a 1970s West Coast style.

Monster 821

The midsize Monster 821 received a styling update for 2018: a more streamlined tank and tail, a new silencer, and a round headlight reminiscent of the original 1993 Monster.

The Euro 4 Testastretta 11° engine delivered slightly less power than before, since the engine acted as a load-bearing element with the frame attaching directly to the cylinder heads. The redesigned rear subframe also attached directly to

SCRAMBLER 1100, 1100 SPECIAL, 1100 SPORT Differing from the Multistrada 1200 Enduro

Engine	Two-cylinder, 4-stroke, air-/oil-cooled
Bore (mm)	98
Stroke (mm)	71.5
Displacement (cc)	1079
Compression ratio	11:1
Valve actuation	Desmodromic single overhead camshaft, 2 valves per cylinder
Injection	Electronic 55mm throttle bodies
Power	86 horsepower at 7,500 rpm
Gears	Six-speed
Front suspension	45mm Marzocchi 48mm Öhlins (Sport)
Rear suspension	Kayaba Öhlins (Sport)
Front brake (mm)	320 dual disc
Rear brake (mm)	245 disc
Wheels	3.50x18, 5.50x17
Tires	120/80x18, 180/55x17
Wheelbase (mm)	1,514
Dry weight (kg)	189 194 (Speciale)

MONSTER 821 2018
Differing from the Monster 821 2017

Power	109 horsepower at 9,250 rpm
Rear tire	180/55x17
Dry weight (kg)	180.5

the engine and supported the new passenger footpeg mounts.

Another update was the color TFT display, and accessories included the Ducati Quick Shift up/down system.

2019

Ducati delivered 53,183 bikes to customers in ninety countries during 2019. Italy remained the number one market, in a generally declining worldwide market. The world's best-selling Superbike for the second year running, the Panigale enjoyed sales totaling 8,304, while two new models introduced in 2019—the Hypermotard 950 and Diavel 1260—managed to sell 4,472 and 3,129 units, respectively. Due to the addition of the 950S and an updated 1260 Enduro, the Multistrada family also performed well: deliveries totaled 12,160. The Scrambler range was boosted with four new models and the Monster with a 25th Anniversary model.

Ducati also entered a prototype V-4 Streetfighter in the Pikes Peak International Hill Climb, with legendary specialist Carlin Dunne setting pole position. Tragically, Dunne died on the way to setting a new course record, crashing in sight of the finish line. When the Panigale V4 25th Anniversary 916 was released at the Laguna Seca WSBK event, bike #5 (Carlin Dunne's racing number) was auctioned to raise funds to support Dunne's family.

In MotoGP the factory Desmosedici GP19 retained its compact V-4 with a counter-rotating crankshaft and large external flywheel. The power was around 290 horsepower at 18,500 rpm. Along with a new aluminum frame, the carbon-fiber swingarm featured thinner and deeper side sections. Aerodynamic updates included a new fairing and wings, plus an aero device incorporated on the swingarm.

Danilo Petrucci replaced Lorenzo for 2019 and joined an elite list of Ducati MotoGP winners. *Ducati*

Although he only managed two race wins, Andrea Dovizioso still finished second overall in the World Championship. Replacing Lorenzo in the factory Ducati team, Danilo Petrucci rewarded Ducati with a victory in his home race at Mugello. Jack Miller and Francesco Bagnaia headed the Pramac Team lineup. Miller rode the latest spec GP19 and achieved five podiums to finish eighth overall in the championship. The final Ducati team was the independent Spanish-based Reale Avintia with riders Tito Rabat and Karel Abraham on earlier spec GP18s.

After thirty-one years of V-twin–powered success in the World Superbike Championship, Ducati introduced the Panigale V4R for 2019. A MotoGP-type Marelli MLE ECU took care of engine management was accomplished, and a titanium Akrapovič 4-2 exhaust system incorporating an exhaust valve handled spent gases. Pierobon provided a racing single-sided swingarm, while the suspension included an Öhlins 46mm RVP2530 fork and RSP40 rear shock absorber. Brembo supplied the P4x30-34 front calipers and

336mm floating discs with a P2x24 caliper and 230mm disc on the rear. MotoGP-style aerodynamic winglets, homologated on the production Panigale V4R, were included for the first time.

Chaz Davies and ex-MotoGP rider Alvaro Bautista handled riding duties. At first the Panigale V4R was successful, Bautista stamping his authority by dominating the first four rounds before his season fell apart. He ultimately won sixteen races and finished second overall. For his part, Davies struggled with the Panigale V4R setup, winning only one race to finish sixth overall.

The BeWiser Ducati pair of Scott Redding and Josh Brookes on Panigale V4Rs, meanwhile, dominated the Bennetts British Superbike Championship, with former MotoGP rider Redding eventually prevailing.

Panigale V4R, V4S Corse, V4 25th Anniversary 916

The Panigale V4R appeared as a homologation model for Ducati's commitment to the V-4 in Superbike racing. To comply with the WSBK 1,000cc limit, the engine featured a 1.1-kilogram-lighter shorter stroke backwards-rotating crankshaft. The 34mm intake valves and con rods were titanium, which, with the lighter internals, allowed the 234-horsepower engine to rev higher and more quickly.

As on the MotoGP Desmosedici, the V4R employed a dry clutch and an Akrapovič racing exhaust. The front frame was optimized for stiffness and the die-cast aluminum swingarm offered a choice of four swingarm pivot positions. The suspension included a race-grade Öhlins NPX 25-30 43mm pressurized front fork and the fairing incorporated MotoGP-style aerodynamic wings. Unlike the Superleggera, the V4R wasn't a limited edition: first-year production was anticipated at one thousand units.

The company brought out an additional version of the Panigale this year: the V4 25th Anniversary 916. While ostensibly a Panigale V4S, the Anniversary 916 included a V4R Ducati Corse racing-style front frame, a dry clutch, and even more track-specific electronics, such as Ducati Quick Shift EVO 2 and predictive DTC EVO 2. Forged magnesium

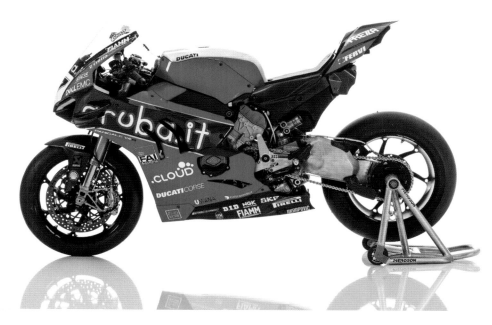

ABOVE: The World Superbike Panigale V4R combined a street bike chassis with a 2015-spec MotoGP engine. Alvaro Bautista used the V4R to good effect, stunning the field with eleven straight race wins at the beginning of the 2019 World Superbike Championship. *Ducati*

BELOW: The Panigale V4R was the most powerful production Ducati yet. MotoGP inspired winglets were incorporated on the fairing. *Ducati*

PANIGALE V4R, 25TH ANNIVERSARY 916
Differing from the Panigale V4S

Stroke (mm)	48.4 (V4R)
Displacement (cc)	998 (V4R)
Injection	Electronic 56mm throttle bodies (V4R)
Power	221 horsepower at 15,250 rpm (V4R)
Wheelbase (mm)	1,471 (V4R)
Dry weight (kg)	172 (V4R) 173 (916 25th)

HYPERMOTARD 950, 950SP
Differing from the Hypermotard 939

Compression ratio	13.3:1
Power	114 horsepower at 9,000 rpm
Front suspension	45mm Marzocchi 48mm Öhlins (SP)
Dry weight (kg)	178 176 (SP)

The Hypermotard 950's minimalist styling cues were reminiscent of the original Hypermotard 1100. The 950SP here also included longer travel Öhlins suspension. *Ducati*

DIAVEL 1260, 1260S
Differing from the Diavel 2018

Stroke (mm)	71.5
Displacement (cc)	1,262
Compression ratio	13:1
Injection	Bosch electronic 56mm throttle bodies
Power	159 horsepower at 9,500 rpm
Front suspension	48mm Öhlins (S)
Rear suspension	Öhlins (S)
Wheelbase (mm)	1,600
Dry weight (kg)	218

Marchesini Racing wheels, a titanium Akrapovič exhaust, and a long list of carbon-fiber and billet aluminum components rounded off this motorcycle's distinctive profile. The styling mirrored the 1999 World Superbike Championship-winning 996; each of the five hundred examples featured an authenticity certificate and a laser-engraved ID number on the top triple clamp.

Hypermotard 950, 950SP

As a replacement for the short-lived Hypermotard 939, a completely updated version appeared in 2019. Racing supermotards influenced its elemental styling, while its wider handlebars and slimmer profile provided more aggressive ergonomics. Minimalist bodywork now exposed the new frame (constructed of tubes of varying diameter); new wheel rims, brake discs with aluminum flanges, and lightweight suspension completed the package.

Although still powered by the 937cc Testastretta 11° engine, the new Hypermotard was 1.5 kilogram lighter and slightly more powerful than before. The electronics package included a Bosch 6-axis IMU to detect roll, yaw, and pitch angles.

The higher-performance Hypermotard 950 SP included dedicated graphics, a flat seat, longer travel Öhlins suspension, and Marchesini forged wheels.

Diavel 1260, 1260S

When it was introduced for the 2011 model year, the Diavel created a new niche market segment. Little had changed in the subsequent eight years, so by 2019 it needed a complete redesign. The bike's new engine, chassis, and electronics package helped this second-generation Diavel 1260 stay faithful to the original, maintaining key styling elements but bringing them up to date.

A Testastretta DVT 1262 engine with variable valve timing, installed in a new chassis, power the new Diavel. Sharper rake and trail aided steering and agility, while a new engine position provided better weight distribution and allowed radiator repositioning to improve cooling.

The sportier S model offered additional features, such as a fully adjustable Öhlins suspension at both front and rear, dedicated wheels, and a higher-performance braking

A Testastretta 1260cc DVT engine powered the second generation Diavel. It was also available as this S version, with Öhlins suspension. *Ducati*

system that included front Brembo M50 calipers. The Diavel 1260 was also compatible with the Ducati Link App, with personalized riding modes, ABS, and traction control and other parameters controlled by a smartphone.

Multistrada 950, 950S, and 1260 Enduro

A new 950 and a higher-specification 950S updated the Multistrada range in 2019, while the Enduro received the 1260 engine. The 950's 937cc Testastretta engine was unchanged, but a new, lighter, double-sided aluminum swingarm was linked to the tubular steel frame. The wheels, both cast alloy and a spoked-wheel option, were lighter, with styling influenced by the larger Multistrada 1260.

The 950S included the electronic Ducati Skyhook Suspension system that instantaneously adjusted fork and shock absorber damping to changes in road surface. Ducati's design for this smallest Multistrada provided the excitement of the larger version in an accessible and versatile package.

The Multistrada Enduro saw an enhancement with the inclusion of the new 1262cc Testastretta DVT engine, accompanied by major chassis and electronics upgrades. The seat, handlebar, and center of gravity were all lower than on the 1200 version, and a new suspension setup improved comfort.

For 2019 the Multistrada Enduro gained a 1260cc engine with updated chassis and electronics. *Ducati*

Scrambler Icon, Full Throttle, Desert Sled, and Café Racer

Four new editions of the Scrambler expanded this already successful platform. Joining the basic Icon were the Flat Track racing-inspired Full Throttle, the 1960s style Café Racer, and the Southern California Desert Sled.

New aesthetic and technical features included more solid aluminum side panels to match the steel teardrop tank, a new headlight, black-painted crankcases, and brushed cylinder head fins. Cornering ABS was added to the electronics suite.

The Full Throttle took its cue from the flat track Scrambler ridden by California racer Frankie Garcia in the 2018

SCRAMBLER ICON, FULL THROTTLE, DESERT SLED, CAFÉ RACER Differing from 2018

Power	73 horsepower at 8,250 rpm
Dry weight (kg)	173 (Icon, Full Throttle)
	193 (Desert Sled)
	180 (Café Racer)

The 2019 Scrambler range. From the left is the Desert Sled, Café Racer, Full Throttle, and Icon. *Ducati*

American Super Hooligan Championship. It also featured a street-homologated Termignoni dual-tailpipe exhaust, low-slung tapered handlebar, and minimalist front fender.

The Café Racer continued the style of the previous version, the new colors of silver with a blue frame recalling 1958's 125GP Desmo.

2020

Toward the end of the previous year, Ducati released six new models and three updated versions for 2020. The COVID 19 pandemic in Italy halted production for seven weeks in April and May 2020. This break affected the first six months of sales, though in the second six months of the year Ducati recorded its best-ever six-month period.

Sales for 2020 ultimately totaled 48,042, headed by the Scrambler with 9,265 units delivered. While Italy remained the premier market, the big mover was China: here sales were up 26 percent (to 4,041), making China Ducati's fourth largest market. The Hypermotard 950 RVE appeared during this year, now featuring street-art–inspired "Graffiti" livery.

Andrea Dovizioso and Danilo Petrucci again rode the factory Desmosedicis in MotoGP. Rising stars Jack Miller and 2018 Moto2 World Champion Francesco Bagnaia were signed for Pramac Racing, with Miller riding a GP20, while 2016 Moto2 World Champion Johann Zarco and Tito Rabat fronted on GP19s with the improving Esponsorama Racing Ducati team.

Ducati continued to lead in innovation. After being the first to introduce a carbon-fiber swingarm and aero wings, Ducati introduced a "holeshot" squatting device that collapsed the rear shock and locked the front fork for race starts. At the same time, the V-4 GP20 struggled with its new-construction Michelin rear tire; intended to encourage smoother cornering arcs, this addition led to inconsistent results. After averaging five victories a year for the previous few seasons, Ducati only managed two during 2020.

Disruptions in the MotoGP season saw the first race delayed until mid-July. Dovizioso began strongly with a victory in Austria, but his season deteriorated overall. Although Petrucci won in France, his season was also unmemorable. The same could be said for Bagnaia. By the end of the season, only Miller remained as a Ducati rider fighting for the podium. With five riders sharing seven podium positions, however, Ducati won their first MotoGP Constructors' World Championship since 2007.

Chaz Davies welcomed 2019 British Superbike Champion Scott Redding as a teammate on the Aruba.it Racing Panigale V4R.

The engine began the season with a 16,100-rpm rev ceiling. It produced a claimed 235 horsepower at the crank, but in other respects it matched 2019's specification.

Redding, Davies, and Michael Ruben Rinaldi won seven races in the shortened season, with Redding and Davies finishing second and third overall. This was the last for Davies: after seven seasons with the factory World Superbike team, Davies' contract wasn't renewed for 2021.

In the truncated Bennetts British Superbike Championship, Christian Iddon teamed with Josh Brookes in Paul Bird's VisionTrack Ducati team. With five race wins, Brookes took his second BSB title.

Panigale V4 Superleggera

The Superleggera V4 appeared in 2020. The third in the Superleggera series, its release was limited to five hundred numbered examples. It was the only street-legal motorcycle available with a carbon-fiber chassis, swingarm, and wheels.

The 998cc Desmosedici Stradale R engine was 2.8 kilograms lighter and even more powerful than the 1,103cc V-4. The racing kit's titanium Akrapovič exhaust helped increase the power to 234 horsepower at 15,500 rpm. Desmosedici GP16-derived biplane wings augmented the carbon-fiber fairing and provided 50 kilograms of downforce at 270 kilometers per hour, 20 kilograms more than that generated by the wings on the newest Panigale V4 and V4R.

In addition, the motorcycle featured a track-type RaceGP display similar to that of the Desmosedici GP20. And thanks to extensive use of carbon fiber, titanium, and machined-

LEFT: Dovizioso won the 2020 Austrian GP, but this was his final season with Ducati. *Ducati*

RIGHT: Scott Redding (left) joined Chaz Davies on the 2020 Aruba.it Ducati Panigale V4R. With four race wins, ex-MotoGP rider Redding was Ducati's best performer in the 2020 World Superbike Championship. *Ducati*

SUPERLEGGERA V4
Differing from the Panigale V4R

Power	224 horsepower at 15,250 rpm
Wheelbase (mm)	1,480
Dry weight (kg)	159, 152.2 (Racing Kit)

Incorporating significant MotoGP technology, the Superleggera V4 was the most exclusive of the new-generation Panigale V4 models. It was the first production motorcycle to offer a carbon-fiber frame, swingarm, and wheels. *Ducati*

A NEW LEASE ON LIFE | **265**

from-solid aluminum components, the dry weight was 16 kilograms less than the Panigale V4, offering a record power to weight ratio of 1.41 horsepower per kilogram.

The Superleggera V4 also sported unique high-level equipment such as a lightened Öhlins suspension system with a pressurized front fork and a Brembo braking system with Stylema R calipers. The electronics package included five additional personalized riding modes. Dedicated Dainese D-air leathers and an Arai carbon-fiber helmet were exclusive features of this bike.

The Superleggera V4 continued Ducati's tradition of offering leading racing technology in a limited production motorcycle and it was undoubtedly the finest Superbike available in 2020.

Panigale V2, Panigale V4, V4S, V4R

The twin-cylinder Panigale was significantly updated for 2020. Replacing the 959 and now called the Panigale V2, its under-engine silencer and single side-end pipe increased the power of this Euro 5-compliant 955cc engine. It also featured a new, comprehensive electronics package based on the 6-axis inertial platform.

Unlike the Panigale 959, the Panigale V2 included Ducati's trademark single-sided aluminum swingarm. Also featured were a restyled fairing and front end dominated by two large air intakes. The motorcycle's design enhanced its comfort as well, with a new seat and suspension setup.

Racing machines throughout the industry had adopted downforce-generating wings in this period, so it was no surprise to see the Panigale V4 adopt wings of its own for 2020. Along with the wings based on the earlier GP16, the front fairing was 15mm wider on each side, the screen 34mm taller, and the lateral fairing extended 38mm outwards.

Updated electronics brought optimized torque delivery in the lower three gears. Chassis updates saw the adoption of the V4R-type Ducati Corse front frame with "optimized stiffness," while suspension modifications resulted in a higher center of gravity and an increased chain-force angle to provide anti-squat and improved stability under acceleration.

The V4R was unchanged for 2020, but a lower-horsepower version was available in the US, Canada, and Mexico.

Streetfighter V4, V4S

After a five-year hiatus, the Streetfighter returned to Ducati's lineup, this time powered by the Desmosedici Stradale V4. Ostensibly a stripped Panigale V4, the Streetfighter received a high and wider handlebar, LED headlight, biplane wings, and a sophisticated electronics package.

Sporting a V-4 engine and slightly longer aluminum front frame, and surrounded by minimalist angular bodywork, the bike's low, lunging front end emphasized a predatory stance.

The Streetfighter V4S shared the V4's engine and chassis but came with Öhlins suspension and forged Marchesini wheels. Offering an unrivaled power to weight ratio in the hyper-naked category, the Streetfighter V4 was also extremely popular, selling 5,730 units during 2020.

Multistrada 1260S Grand Tour

With the Multistrada range about to be replaced, only one new Multistrada appeared in 2020: the 1260S Grand Tour. Designed to provide maximum touring comfort, the Grand Tour was

PANIGALE V2, V4R
Differing from 2019

Power	155 horsepower at 10,750 rpm (V2)
	209 horsepower at 13,250 rpm (V4R North America)
Wheelbase (mm)	1,436 (V2)

Now with updated styling and a single-sided swingarm, the Panigale V2 was the only twin-cylinder Panigale available for 2020. *Ducati*

Offering Superbike performance in a usable package, the Streetfighter V4S successfully established a new niche market for Ducati. *Ducati*

STREETFIGHTER V4, V4S
Differing from the Panigale V4, V4S

Power	208 horsepower at 12,750 rpm
Wheelbase (mm)	1,488
Dry weight (kg)	180, 178 (S)

MULTISTRADA 1260S GRAND TOUR
Differing from the Multistrada 1260S

Dry weight (kg)	215

available in unique Sandstone Gray and red colors. It also came standard with panniers, center stand, dedicated seat, heated handgrips, and a hands-free gas cap that could be opened without a key.

In August 2020 Ducati test rider Andrea Rossi rode a Multistrada 1260 Enduro in the off-road Transanatolia Rally in Turkey, taking first place in the twin-cylinder category and ninth overall.

Scrambler 1100 PRO, 1100 Sport PRO

The successful Scrambler range expanded in 2020 to include two higher-end 1100s: the 1100 PRO and the 1100 Sport PRO. Both shared the Euro 5-spec 1,079cc air-cooled V-twin, now with a double-stacked exhaust and revised fueling. New styling emphasized a more retro look, with a separate number plate holder, shorter rear fender and a round, framed 1970s-style headlight.

Style and equipment differentiated the two 1100 PRP models: the 1100 PRO with the Marzocchi/Kayaba suspension of the previous 1100 Scrambler, forward-mounted footpegs, and a high handlebar. The matt black 1100 Sport PRO featured more aggressive ergonomics, a narrower, shorter handlebar, café racer-style bar-end mirrors, and Öhlins suspension.

In June 2020 a Scrambler 1100 Sport PRO limited edition was built for the exclusive Scuderia Club Italia. Following the previous 1995 Monster 900 Club Italia joint design with Scuderia Club Italia, appearing with a red leather seat by Poltrona Frau, unique graphics, and a range of bespoke aluminum components. With each purchase 500 euros were donated to Sant'Orsola Hospital in Bologna for the fight against the coronavirus.

The only other update to the Scrambler lineup this year was the introduction of the all-black Scrambler Icon Dark.

2021

New models for 2021 included the Panigale V4 SP, Multistrada V4, Monster, SuperSport 950, Diavel Lamborghini, XDiavel Dark and Black Star, and Ducati Scrambler Nightshift and Fasthouse. Although a scarcity of raw materials and lockdowns in various parts of the world led to production challenges, Ducati delivered 49,693 motorcycles during the first nine months

With no engine development allowed, the GP21 was very similar to the GP20. #63 is Francesco Bagnaia and #43 is Jack Miller. *Ducati*

of 2021. This exceeded total 2020 sales and the third quarter of 2021 was the strongest in Ducati's history. The most popular models were the Multistrada V4, Scrambler Nightsift, and Streetfighter V4.

The factory MotoGP rider lineup was new for 2021, with 2020 Pramac riders Jack Miller and Francesco Bagnaia moving up to the official Ducati Corse team. Johann Zarco and 2018 Moto3 World Champion Jorge Martin were now with Pramac Racing, while 2020 Moto2 Champion Enea Bastianini and 2020 Moto2 runner-up Luca Marini rode for the Avintia Esponsorama Racing team.

With engine development frozen this year, the GP21 was ostensibly a stopgap before an all-new engine and chassis appeared for 2022. The GP21 looked similar to the GP20: the box housing the mass damper at the back of the bike received a slight modification and an innovative aero package included new lower side fairing ducts and a slightly different front air intake. This improved straight-line speed and assisted turning. The start device was further developed and, with the GP21 maximizing acceleration, braking, and top speed.

Both the factory Ducati and Pramac teams managed consistent victories and podium finishes. Miller achieved back-to-back victories at Jerez and Le Mans, rookie Martin won in Austria and Bagnaia at Aragón and San Marino. Bagnaia's form improved towards the end of the season with five consecutive pole positions and victories at Portimão and Valencia. He finished second overall in the World Championship with Miller fourth and Zarco fifth. Ducati again took the Constructors' Championship.

In the World Superbike Championship, Scott Redding continued with the factory Aruba.It Racing team while Michael Ruben Rinaldi replaced Chaz Davies. The Panigale V4R was little changed from 2020, with a stated 250 horsepower at 16,000 rpm and a dry weight of 168 kilograms. Both Redding and Rinaldi claimed victories, but a lack of consistency resulted in Redding finishing third and Rinaldi fifth overall.

Panigale V4 SP, V4, V4S

The Panigale V4 family grew with the introduction of the numbered Panigale V4 SP. SP-identified racing versions developed for the Sport Production Championship, a precursor of the current Superstock. The Panigale V4 SP continued this hypothesis: along with a dedicated color scheme and improved chassis components, it provided enhanced racetrack ability. The SP drivetrain featured an STM EVO-SBK dry clutch and lighter 520-pitch drive chain.

Other features set the SP apart, such as the lighter 5-spoke carbon wheels, Brembo Stylema R front brake

The Panigale V4 SP color scheme was inspired by the Ducati Corse "Winter Test" colors for both MotoGP and WSBK. The matte black fairing, carbon rims, and wings contrasted with the exposed brushed aluminum tank. *Ducati*

PANIGALE V4 SP
Differing from the Panigale V4S

Dry weight (kg)	173

MULTISTRADA V4

Engine	Four-cylinder, 4-stroke, liquid-cooled
Bore (mm)	83
Stroke (mm)	53.5
Displacement (cc)	1,158
Compression ratio	14:1
Valve actuation	Valve spring, double overhead camshafts, 4 valves per cylinder
Injection	Electronic 46mm throttle bodies
Power	170 horsepower at 10,500 rpm
Gears	Six-speed
Front suspension	50mm Marzocchi
Rear suspension	Marzocchi
Front brake (mm)	330 dual disc
Rear brake (mm)	265 disc
Wheels	3.00x19, 4.50x17
Tires	120/70x19, 170/60x17
Wheelbase (mm)	1,567
Dry weight (kg)	215, 218 (S), 217 (Sport)

MONSTER 2021
Differing from the Monster 821

Bore (mm)	94
Displacement (cc)	937
Compression ratio	13.3:1
Power	111 horsepower at 9,250 rpm
Wheelbase (mm)	1,474
Dry weight (kg)	166

SUPERSPORT 950, 950S
Differing from the SuperSport

Power	110 horsepower at 9,000 rpm

calipers, and a Brembo front brake radial master cylinder with a Multiple Click System (MCS) that allowed instant brake feel adjustment. Billet aluminum adjustable footpegs, a carbon-fiber front fender, and open clutch cover completed the package of SP equipment.

Intended as a track bike for the amateur rider, the SP was more user-friendly than the more demanding Panigale V4R. For 2021 the Panigale V4 and V4 S were now Euro 5 compliant, albeit with maximum power and torque unchanged. The electronics package continued to evolve, now with more predictive DTC and new riding modes.

Multistrada V4, V4S, V4S Sport

Since its release in 2003, the Multistrada had evolved three times. The first Multistrada was primarily a street-oriented sportbike, embodying the "four bikes in one" concept in 2010. This offered more performance, greater comfort, and improved aerodynamics.

In 2015 the Multistrada gained the first variable-timing Testastretta, while the fourth-generation model received a new V-4 engine. This was the first Ducati engine in thirty years to include valve springs, eschewing the trademark Desmodromic valve actuation system.

The compact V4 Granturismo engine was similar in design to the Desmosedici Stradale, but a slightly bigger bore provided a displacement increase and the engine was tuned for a more linear torque curve. The counter-rotating crankshaft featured 70-degree offset crankpins with a twin-pulse ignition that offered the sound and feel of a twin-cylinder. The valve-spring cylinder head layout allowed for 60,000-kilometer valve adjustment intervals. An aluminum monocoque frame replaced the previous Multistrada's tubular steel trellis type, and the swingarm was now double sided. The front wheel was 19 inches and spoked wheels were also available on the S and Sport.

Revised ergonomics saw the rider more centrally located to improve low-speed maneuverability, and the updated aerodynamics reduced noise and improved comfort. As expected, the electronics package was state of the art, where the inertial platform managed multiple functions, including radar technology, a first for a production motorcycle.

The S and Sport also received electronic Skyhook suspension. Optional accessory packs enhanced performance, enduro, touring, or urban ability. An immediate success, the Multistrada V4 sold 5,000 units within the first six months of production. The existing Multistrada 960, 950S, and 1260 Enduro continued as before, with the 950S now available in GP White.

Monster

Although updated in 2018, a significantly new Monster appeared in 2021. While maintaining the compact, minimalist style that had typified the Monster since 1993, the Testastretta 11° engine now provided more displacement, power, and torque compared to the previous 821. A lighter Panigale V4-inspired short aluminum front frame attached directly to the cylinder heads. Along with lighter wheels, aluminum swingarm, and a Glass Fiber Reinforced Polymer rear subframe, the new Monster was 18 kilograms lighter than before.

The styling evoked a direct link to the previous generation Monster, with the headlight recalling the round shape of earlier incarnations. Designed to appeal to a wider range of riders, the seat was low and narrow, with a comprehensive electronics kit included as standard. The Monster was also available as a Plus version, featuring a small windscreen and passenger seat cover.

The Monster 1200 and 1200S were the only other Monsters offered for 2021.

SuperSport 950, 950S

Four years after its initial launch, the SuperSport had evolved into the SuperSport 950. It was now Euro 5 compliant, featuring new electronics, but its most visible update was new styling. This model offered a fairing design similar to the Panigale V4, with air vents incorporated on the fairing sides and the lower fairing now extending to the silencer. The user-friendly style hadn't changed, and the SuperSport 950 continued as a practical all-round sportbike.

Diavel 1260 Lamborghini

With both Ducati and Lamborghini now under the Audi umbrella, a collaboration was inevitable. Based on the Diavel 1260 S, and influenced by the hybrid Lamborghini Siàn FKP 37, the Diavel Lamborghini transferred some basic concepts of the Siàn FKP 37 design to the Diavel.

TOP: The 2021 Monster was more powerful and compact, and significantly lighter than before. *Ducati*

MIDDLE: For 2021, the SuperSport evolved into the SuperSport 950. This had a new fairing and the 950S had Öhlins suspension. *Ducati*

BOTTOM: The Diavel 1260 Lamborghini was a limited-edition model inspired by the Lamborghini Siàn FKP 37. *Ducati*

DIAVEL 1260 LAMBORGHINI
Differing from the Diavel 1260S

Dry weight (kg)	220

XDIAVEL DARK, S, BLACK STAR
Differing from the XDiavel 2020

Power	160 horsepower at 9,500 rpm
Dry weight (kg)	221 (Dark, Black Star), 223 (S)

Featuring new automotive-inspired forged wheels, the 1260 S also offered carbon-fiber details, including radiator covers, air intakes, fenders, various covers, and the headlight frame. The paint was the same as that used for the Siàn FKP 37, with bodywork in Gea Green and the frame, undertray, and forged rims in Electrum Gold. The number 63 signified Lamborghini's founding year. Although the Siàn was limited to only sixty-three units, the Diavel 1260 Lamborghini saw a total production of 630 numbered examples.

XDiavel Dark and Black Star

The XDiavel lineup grew to three models for 2021, the XDiavel Dark and Black Star joining the sporting XDiavelS. The Euro 5 spec engine included a new exhaust system and produced more power than before. The dual-outlet exhaust now exited rearward.

Intended as a gateway to the feet-forward XDiavel world, the basic-specification Dark continued the matte-black style that had begun in 1998 with the Monster 600 Dark. The range-leading XDiavel Black Star was inspired by the world of sports cars, with a color scheme in dedicated grey and matte black set off with red cylinder head covers.

The XDiavel Black Star also included lighter forged and machined wheels, and both the S and Black Star featured improved Brembo M50 front brake calipers with a PR16/19 radial master cylinder. Sport, Touring, and Urban personalization packages were also available. With the X of XDiavel representing Crossover, the XDiavel continued to successfully synthesize two distinct spectrums: the traditional, relaxed low-speed cruiser and Ducati's natural, adrenaline-charged sportiness.

Scrambler Nightshift, Fasthouse

The Scrambler range was simplified slightly for 2021, with the Nightshift replacing the Café Racer and Full Throttle variants. This year saw the introduction of new versions of the Desert Sled and 1100 Dark PRO, along with a limited-edition Fasthouse.

The Nightshift emphasized a classic custom style, with a straight, narrow handlebar and café racer bar-end mirrors. Side number plates, spoked wheels, a minimal front fender, no rear fender, and a grey-and-black color scheme contributed to the Scrambler Nightshift's seemingly bespoke character.

The limited-edition Scrambler Fasthouse celebrated Ducati's success in the 2020 Mint 400 off-road race. *Ducati*

The colors of the Panigale V2 Bayliss "1st Championship 20th Anniversary" replicated Troy Bayliss' 2001 996R. *Ducati*

Following Jordan Graham's victory in the Hooligan class of the 2020 Mint 400—the oldest and most prestigious off-road race in America—Ducati released a limited-edition Scrambler in collaboration with the clothing brand Fasthouse. Based on the Desert Sled, the Fasthouse was distinguished by special livery, reflecting the race bikes' look, and a custom seat. Each of the eight hundred examples received a numbered plate.

2022

The off-road DesertX headed the 2022 lineup, and apart from an updated Panigale V4, most other new models centered on expanding the twin-cylinder V2 range. In MotoGP Ducati predominated the grid, with eight Desmosedicis entered. Jack Miller and Francesco Bagnaia continued with the factory Ducati Lenovo Team, as did Johann Zarco and Jorge Martin for Pramac Racing. Gresini Racing switched from Aprilia, signing a two-year deal with Ducati with Enea Bastianini and Fabio Di Giannantonio on 2021-spec machines. The new VR46 Racing Team also signed a three-year contract with Ducati, their two bikes ridden by Luca Marini and Marco Bezzecchi.

In the factory Aruba.it Racing Ducati team for the World Superbike Championship, Alvaro Bautista returned alongside Michael Ruben Rinaldi. A change in World Supersport regulations for 2022 allowed larger capacity twins and triples. Ducati announced a factory-backed Panigale V2 to be ridden by ex-Moto3 and Moto2 rider Nicolo Bulega. The Barni Racing team also entered Panigale V2s for twenty-year-old Luca Bernardi and eighteen-year-old Oli Bayliss, son of former racer Troy Bayliss.

Panigale V2 Bayliss "1st Championship 20th Anniversary"

The Panigale V2 Bayliss celebrated the twentieth anniversary of Troy Bayliss' first World Superbike title. This model's celebratory livery was similar to Bayliss' 2001 Ducati 996R. Based on the Panigale V2, the numbered Bayliss edition featured Öhlins suspension and steering damper, with optional solo seat. Other special equipment included sports grips, carbon fiber and titanium silencer outlet, and a double, red-stitched seat.

V2 PANIGALE BAYLISS 20TH ANNIVERSARY Differing from the Panigale V2	
Front suspension	43mm Öhlins NIX30
Rear suspension	Öhlins TTX36
Wheelbase (mm)	1,438
Dry weight (kg)	174.5
Dry weight (kg)	166

MULTISTRADA V4 PIKES PEAK, V2, V2S
Differing from the Multistrada V4 and 950

Front suspension	48mm Öhlins (Pikes Peak)
Rear suspension	Öhlins TTX36 (Pikes Peak)
Wheels	3.50x17 and 6.0x17 (Pikes Peak)
Tires	120/70x17 and 190/55x17 (Pikes Peak)
Wheelbase (mm)	1,595 (Pikes Peak)
Dry weight (kg)	214 (Pikes Peak), 199 (V2), 202 (V2S)

Multistrada V4 Pikes Peak, V2, V2S

The Multistrada V4 Pikes Peak was the sportiest Multistrada yet, featuring an Akrapovič titanium and carbon silencer and Race Riding Mode option. The racing emphasis was reflected in an aluminum monocoque frame with revised steering geometry, a 17-inch front wheel, and a single-sided swingarm. The wheels were light forged-aluminum Marchesini and its Öhlins Smart EC 2.0 suspension included an "event based" mode that automatically adjusted the setting according to the user's riding style. Revised ergonomics featured a lower handlebar and higher and more rearward footpegs, while a range of carbon-fiber components, MotoGP colors and graphics, low smoked Plexiglas screen, gold anodized fork legs, and Ducati Corse badges emphasized the sporting style.

For 2022 the Multistrada V2 replaced the Multistrada 950 as the gateway to Ducati's touring lineup. Updates centered on improved ergonomics, weight reduction, and engine and electronic improvements. This latest incarnation of the 937cc Testastretta 11° engine, included new connecting-rods, eight-disc hydraulic clutch, and an updated gearbox. Several components, including footpegs, wheels, and front brake discs, were shared with the Multistrada V4 and contributed to a 5kg weight reduction from compared the Multistrada 950. The Multistrada V2S included Sachs Skyhook electronic suspension. The Travel trim option for the Multistrada V2S included side bags, heated grips and a center stand.

Hypermotard 950 SP

The Hypermotard range was updated to meet the Euro 5 standard and comprised three models: Hypermotard 950, Hypermotard 950 RVE, and Hypermotard 950 SP. The power and torque were unchanged, but the gearbox was updated

With MotoGP-inspired styling and graphics and a 17-inch front wheel, the Multistrada V4 Pikes Peak was the most sporting of the Multistrada line-up. *Ducati*

with the gearshift drum now rotating in bearings to improve neutral selection. The top-of-the-range Hypermotard SP featured MotoGP-inspired colors, longer travel Öhlins suspension, Marchesini forged wheels, and standard Ducati Quick Shift (DQS) Up and Down EVO.

Diavel 1260S "Black and Steel"

The Diavel 1260 family expanded for 2022 with the introduction of the 1260S "Black and Steel." This version was inspired by the 2019 concept of the Diavel "Materico" presented at the "Beautiful Boldness/Visionary Design" event held during Milan Design Week, and the subsequent display at the MOARD (Motorcycle Art &Design) exhibition. The Diavel 1260S

"Black and Steel" was characterized by asymmetrical graphics that combined glossy gray and matte black with a yellow frame and other highlights.

Scrambler 1100 Tribute PRO, Urban Motard

The Scrambler 1100 Tribute PRO was released to celebrate the fiftieth anniversary of the first 90-degree V-twin Ducati, the 1971 750 GT. The Giallo Ocra colors paid homage to the 1972-74 750 Sport. Classic logos from the Giorgetto Giugiaro era, black spoked wheels, circular rear-view mirrors, and a brown stitched seat further emphasized the 1970s vibe. The 800cc Urban Motard was the other new Scrambler. Billed as a "city rebel," this version was intended for newer, city riders. Street art and metropolitan graffiti inspired the Star White Silk and Ducati GP19 Red graphics while black spoked wheels, high front fender, black flat seat, and side number plates evoked the Motard style.

Streetfighter V2, V4 SP

Two new Streetfighter models were released for 2022: the Panigale V2-based Streetfighter V2 and high-specification Streetfighter V4 SP. The Streetfighter V2 was ostensibly an unfaired Panigale V2 with higher and wider aluminum handlebars, and marketed as gateway to the Streetfighter family. The twin-cylinder 955cc Streetfighter V2 engine incorporated the latest electronics package and produced slightly less power than the Panigale V2. Built around the Panigale V2's mechanicals, the design concept included a wider seat and repositioned footpegs. The Streetfighter V2 retained the engine as a load-bearing component attached to the diecast aluminum monocoque frame. Its single-sided swingarm was 16mm longer than that of the Panigale V2. Brakes, wheels, and suspension were unchanged.

The range-topping Streetfighter V4 SP combined the Streetfighter "Fight Formula" with SP (Sport Production) specifications. Apart from an STM-EVO SBK dry clutch, the V4 Desmostradale was unchanged. Premium Superleggera V4-derived equipment included significantly lighter split-spoke carbon rims. As on the Panigale V4 SP a matte black "Winter Test" Ducati Corse color scheme contrasted with bright red accents and a brushed aluminum tank.

Classic Giugiaro graphics and 1970s colors distinguished the celebratory Scrambler 1100 Tribute PRO. *Ducati*

SCRAMBLER URBAN MOTARD
Differing from the Scrambler Nightshift

Wheelbase (mm)	1,436

STREETFIGHTER V2, V4SP
Differing from the Panigale V2 and Streetfighter V4S

Power	153 horsepower at 10,750rpm (V2)
Wheelbase (mm)	1,465 (V2)
Dry weight (kg)	178 (V2); 196 (V4SP)

The 2022 Streetfighter line-up; from the left the new Streetfighter V2, Streetfighter V4 SP and Streetfighter V4. *Ducati*

Leading factory MotoGP rider Francesco "Pecco" Bagnaia poses with the 2022 Panigale V4S. *Ducati*

Panigale V4, V4S

To improve track performance, the Panigale V4 was updated significantly for 2022. While the basic engine architecture was unchanged, the lubrication circuit and oil pump were more efficient. The silencer outlets were increased to 38mm to reduce back pressure and increase power. An optional Akrapovič titanium exhaust system reduced the weight by 6kg and increased the power to 228 horsepower. A closer-ratio gearbox featured more race-oriented ratios, with taller first, second, and sixth gears. The result was a top speed increase of 3 mph over the 2021 version.

The basic frame was unchanged, but the swingarm pivot was positioned 4 mm higher to increase anti-squat under acceleration. The suspension was also new for the Panigale V4S with the Öhlins fork including a pressurized cartridge damping system similar to the Öhlins racing forks. A 25mm piston managed compression in the left leg while a 30mm piston controlled rebound damping in the right leg. Aerodynamic updates resulted in a new fairing with thinner double-profile wings and revised extraction air ducts to improve cooling. Both the saddle and fuel tank were redesigned to improve rider ergonomics, while the graphics and latest generation TFT dashboard were also new.

DesertX

Initially presented as a concept bike in November 2019, the DesertX is the most off-road focused motorcycle ever offered by Ducati. It's powered by the same 937cc liquid-cooled Testastretta 11° Desmodromic twin as used in the Monster and Multistrada V2, but its six-speed gearbox features more

The twin headlight styling of the Cagiva Paris-Dakar racers of the early 1990s inspired the DesertX. *Ducati*

off-road suited ratios. Electronic aids include six riding modes in combination with four power modes. A new steel trellis frame is supported by a long-travel suspension and 21-inch front and 18-inch rear wheels. The slim profile was designed to provide both control in a standing position and road comfort, while a Plexiglas fairing similar in style to the original Paris-Dakar Cagivas offers aerodynamic protection.

In a world moving toward electric vehicles, it was inevitable Ducati would embrace this technology at some stage. At the end of 2021 Ducati signed an agreement with Dorna Sports to be the sole official supplier of motorcycles for the FIM Enel MotoE™ World Cup from 2023 until 2026. MotoE™ will be the electric class of the MotoGP World Championship. This move continued Ducati's long tradition of using racing and competition as a blueprint for future production bikes. This approach began with Ing. Fabio Taglioni's first Gran Sport in the 1950s and continued with the round-case bevel drive 750 Super Sport of 1974 the 851 and 916 Desmoquattros through to the latest Panigale V4. Ducati's reputation has always been built on racing success and production race replicas, and there is no suggestion this will change even as new technologies loom.

PANIGALE V4, V4S
Differing from 2021

Power	215.5 horsepower at 13,000rpm 210 horsepower at 12,000rpm (USA)
Front suspension	Öhlins 43mm NPX25/30 (V4S)

DESERTX
Differing from the Multistrada V2

Compression ratio	13.3:1
Power	110 horsepower at 9,250rpm
Front suspension	46mm Kayaba
Rear suspension	Kayaba Monoshock
Front brake (mm)	320 dual disc
Rear brake (mm)	265 disc
Wheels	2.15x21, 4.50x18
Tires	90/90x21, 150/70x18
Wheelbase (mm)	1,608

A NEW LEASE ON LIFE

INDEX

48, 16
48 Brio, 75, 76
48 Brisk, 66-67
48 Cacciatore, 69-70, 88-89
48 Piuma, 66-67, 69-70, 71
48 Piuma Sport, 69-70
48 Sport, 46, 66-67, 69-70
48 Sport Export, 70
48/1 Brisk, 69-70
48SL, 69-70, 88-89
50 Falcon, 69-70, 75
50 Piuma, 89
50 Piuma Sport, 83
50 Scrambler, 96-97, 107, 115
50SL, 83-84, 88-89
50SL/1, 88-89
50SL1, 96
50SL/1A, 94
50SL/2, 83, 88-89
50SL/2A, 94
55E, 25, 33
55R, 25, 33
60, 12, 13
    Sport, 14-15
65, Sport, 14-15
65T, 15, 24, 33, 38, 46
65TL, 15, 24, 38, 46
65TS, 33, 38, 46
65TS (Turismo Sport), 24
80 Falcon, 69-70
80 Setter, 69-70, 75, 80
80 Sport, 69-70, 75, 80
85 Bronco, 52, 57, 61, 66
85 Sport, 46, 47, 52, 57, 61
85 Super Sport, 57, 61
85 Turismo, 46
85T, 47, 52, 57
86T, 123
86TE, 123
90 Cacciatore, 75
90 Cadet, 75, 80, 83
90 Falcon, 75, 80, 83, 88-89
90 Mountaineer, 75, 80, 83, 88-89
98, 18, 25
    Sport, 19-21
    Super Sport, 19-21
98 Bronco, 52, 54, 57, 61
98 Bronco (Cavallino), 66
98 Cavallino, 52, 57
98 Sport, 24, 25
98N, 33-34, 38
98S, 33, 38
98T, 19, 25, 33, 38
98TL, 19, 25, 33-34, 38, 46
98TS, 46, 47, 48, 52, 54, 57, 61, 66, 69
100 Brio, 80
100 Cadet, 80, 83-84, 88-90, 94, 96-97
100 Gran Sport (Marianna), 26-27
100 Mountaineer, 80, 83-84, 88-90, 94, 96
100 Scrambler, 96-97, 107
100 Sport, 43, 45, 52-53
100/25 Brio, 83-84, 88-89, 94, 96
125 Aurea, 61, 69
125 Bronco, 74, 80, 83, 87
125 Cadet/4, 87, 94, 96-97
125 Cadet/4 Lusso, 87, 94, 96-97
125 Cadet/4 Scrambler, 87, 94, 96, 96-97
125 Gran Prix (Bialbero), 29-30
125 Monza, 52, 61
125 Regolarità, 121, 125, 128, 130, 133
125 Scrambler, 107-108, 111
125 Six Days, 130, 133
125 Sport, 43-44, 47, 52-53, 57, 65, 74, 77-78, 81-83, 87
125 Super Sport, 52
125F3, 48-50
125S, 32-33, 38, 46
125T, 38, 46, 52
125TS, 44-45, 52-53, 57, 61, 65-66, 74
125TV, 32-33, 38, 46, 52

160 Monza Junior, 77, 78-79, 81-83, 87, 91-92, 96-97
175 Americano, 42, 52, 56-57
175 Motocross (Scrambler), 43, 52, 56-57, 61
175 Silverstone, 52
175 Sport, 34, 36, 42, 50, 52, 56-57
175 Super Sport, 42, 52
175F3, 48-50, 59
175T, 34, 37-38, 42, 52, 56-57
175TS, 52, 61
200 Americano, 51, 65-66
200 Élite, 50, 56, 61, 62, 65, 74, 77-78, 81-83
200 Motocross (Scrambler), 51, 56, 61
200 Super Sport, 51, 56, 82
200GT, 62, 65, 69, 74, 77-78
200TS, 51
200TS Americano, 56, 61, 62
239 Scrambler, 119
250 Desmo, 103-105, 111, 115-116, 119
250 Desmo Twin, 55-56
250 Diana, 69
250 Diana (Daytona), 59-60, 63-65
250 Diana Mark 3, 62, 63-65, 69
250 Diana/Daytona, 74
250 Mach 1, 71, 72-73, 77-78, 81
250 Mach 1S, 71, 73
250 Mark 3, 71-72, 77-78, 81, 86, 87, 92-93, 96-97, 106, 111, 115-116, 119
250 Mark 3 Desmo, 93, 96-97
250 Monza, 59-61, 63-65, 69, 74, 77, 81-82, 87, 91-92, 96
250 Scrambler, 62-63, 69, 74, 77, 81-83, 87, 90-91, 96, 96-97, 106, 111, 115-116, 119
250F3, 59, 63-65, 71
250GT, 71, 72, 81-83
250R/T, 119
250SC, 76-77, 81
250SCD, 85
300F3, 154
350 Desmo, 103-105, 111, 115-116, 119
350 Desmo Twin, 55-56
350 Indiana, 154-155
350 Mark 3, 92-93, 96-97, 106, 111, 115-116, 119
350 Mark 3 Desmo, 93, 96-97
350 Scrambler, 77, 81-83, 87, 90-91, 96-97, 106, 111, 115-116, 119
350 Sebring, 77-78, 79, 82-83, 87, 91-92
350 Sport Desmo, 129, 133, 139
350 Supersport, 166-167, 171
350 Triple, 102
350GTL, 124-125
350GTV, 129-130, 133
350R/T, 105-106, 119
350SC, 76-77, 81
350SCD, 85
350SL, 144-145
350TL, 144-145
350XL, 143, 144-145
360 SSS, 96
400 Monster, 184, 193
400 Supersport, 164, 166-167, 171, 175-176, 179, 181, 184
400 Supersport Junior, 162
400F3, 154, 156
450 Desmo, 103-105, 111, 115-116, 119
450 Mark 3, 95-96, 96-97, 106, 111, 115-116, 119
450 Mark 3 Desmo, 95-96, 96-97
450 Scrambler, 95-96, 96-97, 106, 111, 115-116, 119
450R/T, 96, 101, 105-106, 111, 119
500 Grand Prix Twin, 101, 102
500 Sport Desmo, 127, 129, 133, 139
500GTL, 124-125
500GTV, 129, 134
500SL Pantah, 134, 136, 137, 139, 140, 143, 144

600 Monster, 175, 179, 181, 184, 188, 190, 193, 197
600 Monster Dark, 188
600 Pantah, 147
600 Supersport, 175-176, 179, 181, 184
600SL Pantah, 139, 140, 143, 144, 176
600TL Pantah, 141, 142-143, 144
600TT2, 141
620, 202
620 I.E. Monster, 197
620 Monster, 201, 204-205, 207, 209-210
620 Multistrada, 207
620 Supersport, 202
620S Capirossi, 204-205
650 Alazzurra, 150
650 Indiana, 154-155
650SL, 144-145
748, 193, 195-196, 200-201
748 Biposto, 181, 182-183, 186, 189
748 Desmoquattro, 176
748 Economy, 191
748 Racing, 186
748 Strada, 177-178, 181, 182-183, 186
748R, 191, 193, 195-196
748S, 191, 193, 195-196
748SP, 177-178, 181, 182-183, 186
748SPS, 189
749, 200-201, 203-204, 206, 209
749R, 203-204, 206
749S, 200-201, 203-204, 206
750 I.E. Monster, 197
750 Indiana, 154-155
750 Monster, 181, 184, 188, 190, 193, 199
750 Paso, 151-152, 159, 161
750 Sport, 101, 109, 110, 111, 113-114, 118, 133, 156, 159, 161, 194, 210
750 Super Sport, 101, 113, 116-117, 123-124, 126-127, 128, 132-133, 164, 210
750 Supersport, 164, 166-167, 171, 175-176, 179, 181, 184, 188, 190, 191-192
750 TT1, 145, 146
750F1, 149-150, 152, 156
750F1 Laguna Seca, 156
750F1 Montjuïc, 152-153
750F1 Santamonica, 158
750F1S, 156
750GT, 101, 102-103, 108, 110, 113-114, 118, 122, 133
800 Monster, 199, 205
800 Supersport, 199, 202, 204
848, 217-218, 219, 223, 226
848 Corse SE, 233-234, 236
848 EVO, 229, 233-234, 236
851 Desmoquattro, 156
851 Racing Replica, 160
851 Roche Replica, 160
851 SP2, 161
851 SP3, 161
851 Strada, 157, 159, 160, 161, 163, 166, 167, 170
851 Superbike Kit, 157
860GT, 121-122, 124
860GTE, 121-122
860GTS, 126, 128
888 Corsa, 163, 167
888 Racing, 166, 168, 173
888 SP4, 166
888 SP5, 169-171
888 SPO, 169-171
888 SPS, 166
888 Strada, 169-171
899 Panigale, 238, 239-241
900 California, 190
900 Cromo, 188, 190
900 I.E., 166-167
900 I.E. Monster, 197
900 Mike Hailwood Replica, 134, 137, 138, 139-140, 142, 143, 144

900 Monster, 167, 175, 179, 181, 183, 184, 188, 190, 192, 193, 199
900 Sport, 196-197
900 Super Sport, 121, 123-124, 125, 126-127, 128, 130, 132-133, 135, 137, 138, 139-141, 142
900 Super Sport Darmah, 134, 135
900 Superlight, 166-167, 171, 175-176, 179, 181
900 Supersport, 161, 162, 164, 166-167, 171, 175-176, 179, 181, 184, 187, 189, 191-192, 196
900CR, 175-176, 179, 181, 185
900GTS, 132-133, 135
900NCR, 131-132, 134, 195
900S, 190
900S2, 142, 143, 144
900SD Darmah, 127, 128-129, 130, 132-133, 135, 137, 138, 139-140, 142
900SP, 175-176, 179, 181, 184
900SS Final Edition, 187
900SSD Darmah, 135, 137, 138, 139-141
906 Paso, 158-159, 161, 164
907 I.E., 164-165, 167, 183
916 Biposto, 181, 182-183, 186
916 Monoposto, 186
916 Racing, 177
916 Senna, 176, 177-178, 182-183, 186
916 Senna 1, 240
916 SP, 173-174
916 Strada, 173-174, 177-178, 181, 182-183, 186
916SP, 175, 177-178
916SP3, 181
916SPS, 182-183, 186
916SPS Fogarty Replica, 186
955 Ferracci, 173
955SP, 181
959 Panigale Corse, 258
996, 193
996 Biposto, 189, 191
996 Desmoquattro, 190, 192, 194
996 Factory Replica 2, 191
996 Pista, 191
996 S4R, 204
996 Strada, 189
996R, 193, 199
996S, 192, 193
996SPS, 189, 191
998, 195-196, 199, 200-201
998 FE, 203-204
998 Matrix, 203-204
998 Testastretta, 195
998F02, 195, 196
998R, 195-196
998S, 195-196
998S Bayliss, 196-197
998S Bostrom, 196-197
999, 199, 200-201, 203-204, 206, 207, 209
999 Fila, 200-201
999F03, 199
999F04 Fila, 203
999F05, 205
999F06, 209
999R, 200-201, 203-204, 206
999R Xerox, 209
999S, 200-201, 203-204, 206
1000, 199
1000 Monster, 202, 204-205, 207, 209
1000 Multistrada, 207
1000 Supersport, 202, 204
1000DS, 202, 204, 210
1000DS Supersport, 209
1098, 213-214, 217-218, 219, 221
1098 Streetfighter, 235
1098F08, 216
1098F09, 222
1098R, 216, 217-218, 219, 223
1098R Bayliss, 223
1098S, 213-214, 217-218, 219
1098S Tricolore, 214

1198, 221, 223, 226, 229
1198F09, 221
1198F10, 225
1198R, 226
1198R Corse Special Edition, 226
1198RS11, 229
1198RS12, 233
1198S, 223, 226
1198SP, 229
1199 Panigale, 233–234, 236
1199 Panigale R, 235, 236, 239–240
1199 Panigale S, 233–234, 236
1199 Panigale S Senna, 238, 241
1199 Panigale Senna, 239–240
1199 Panigale Tricolore, 233–234, 236
1199 Superleggera, 239–240, 241
1299 Panigale, 241
1299 Panigale R Final Edition, 256–257
1299 Panigale S, 244
1299 Panigale S Anniversario, 251–252
1299 Superleggera, 252
2501GT, 71–72

A

Abraham, Karel, 260
Aermacchi, 151
Agostini, Giacomo, 102, 108
Alazzurra, 144
AMA Championship, 165
AMA Superbike National Championship, 168, 170, 173
Amadori, 26, 29, 48, 59
Amal, 81, 103, 107
Apollo, 67–69
Aprilia, 56, 238
APTC, 207
Armaroli, 136
Aruba.it Racing team, 246, 251, 256, 265, 269, 273
Audi AG, 233, 234
Australian Superbike Championship, 188

B

Badovini, Ayrton, 236
Bagnaia, Francesco, 260, 264, 269, 273
Barcelona 24-hour race, 34, 112, 121, 137, 146
Barcone, 54
Basset, Damien, 224
Bastianini, Enea, 269, 273
Battilani, 23
Battle of the Twins, 143, 148, 150, 155, 156
Bautista, Alvaro, 261, 273
Bayliss, Oli, 273
Bayliss, Troy, 188, 192, 195, 199, 208–209, 213, 216, 217–218, 221, 223, 250
Bennetts British Superbike Championship, 261, 265
Berliner, 34, 42, 54, 56, 62, 63, 73, 74, 75, 78, 86, 96, 97, 105, 112, 126
Berliner, Joe, 67, 69
Bezzecchi, Marco, 273
Bialbero, 29, 30–31, 35, 41
Bilancioni, Franco, 136, 157
Bing, 154
Biposto 916, 176, 177–178
BMW S1000R, 248
Bol d'Or d'Italia 24-hour endurance race, 141, 146, 151
Bordi, Massimo, 151
Borgo, 37, 73
Borrani, 105, 106, 115, 132, 135
Bosch, 128, 129, 132, 135, 237, 240, 244, 246
Bosch-Brembo, 226
Bostrom, Ben, 190, 192, 196
Brembo, 118, 119, 132, 156, 160, 161, 162, 163, 164, 165, 169, 171, 201, 227, 234, 239, 248
Brembo Gold, 153
Brio, 71, 75, 76
Brisk 50/1, 88–89
British 125 Championship, 46, 54
British Superbike Championship, 177, 188, 190, 195, 200, 205, 216
Broccoli, Massimo, 139
Brookes, Josh, 261, 265

Bulega, Nicolo, 273
BSA, 108
 441 Victor, 105
 Victor, 95
Byrne, Shane, 200, 216, 251

C

Cagiva, 143, 145, 149, 173
Calcagnile, 98
Calder, Frank, 73
Campagnolo, 128
Canellas, Salvador, 112, 121
Canepa, Niccolò, 214
Capirossi, Loris, 199, 203, 205, 206, 208, 213, 216
Caproni, 13, 15
Carello, 143
Carini, Mario, 32
Castiglioni brothers, 149, 150
Catalunya, 199, 208
Ceccato, 23
Ceriani, 45, 102, 119
Chadwick, Dave, 39
Checa, Carlos, 205, 228–229, 233, 236
Ciceri, Santo, 32
Ciclo scooter, 16
Clymer, Floyd, 74
Conti, 103, 121, 123, 126, 134, 135, 156
Cors, 166
Corser, Troy, 173, 179–180, 185, 188
Cruiser, 16–18
Crutchlow, Cal, 238
Cucciolo, 7, 9, 24, 251
 M55, 25
 racing, 13
 T1, 8, 9–10
 T2, 10, 11
 T2 Sport, 11, 13
 T2 Turismo, 11
 T3, 11, 12, 13
 T50, 11
Cussigh, Walter, 141

D

D16GP4, 203, 205
D16GP5, 205, 206
D16GP6, 208, 213
D16GP7, 213
D16GP8, 216
D16GP9, 221–222
D16GP10, 225
D16GP11, 228–229
D16GP12, 233
D16GP14, 238
D16RR, 218
Dainese, 244
Dall'Inga, Luigi, 238, 242, 247
Daspa, 131
Davies, Chaz, 239, 240, 242, 243, 247, 251, 256, 261, 265, 269
Daytona, 48, 85, 120, 128, 143, 148, 150, 155, 228
De Eccher, Cristiano, 112, 125
Degli Antoni, Gianni, 26, 27, 28, 29, 31
Del Piano, Guido, 137
Del Torchio, Gabriele, 213, 233
Dell'Orto, 13, 16, 28, 30, 35, 37, 40, 55, 62, 63, 67, 73, 78, 93, 95, 96, 104, 108, 110, 117
DesertX, 276–277
Desmodromics, 29, 54, 81, 204
 125, 30–31, 35
 125 Single and Twin, 39–40
Desmodue, 201, 208, 224
Desmoquattro, 151, 157, 173, 177, 189, 192
Desmoquattro Monster S4, 193
Desmosedici, 195, 198, 199, 203, 213, 216, 233, 243, 273
 696, 216
 848, 216
 1098R, 216
 GP12, 232
 GP13, 235
 GP14, 239
 GP17, 251
 RR, 216–217, 218
Diavel, 213, 230, 234–235, 238, 241, 247
 1260, 260, 262

1260 Lamborghini, 271–272
1260S, 262–263
1260S "Black and Steel," 274–275
 AMG Special, 234
 Carbon, 230, 234, 238, 247, 247–248
 Cromo, 234
 Diesel, 254–255
 Strada, 241, 242
 Titanium, 241
Di Giannantonio, Fabio, 273
Digiplex Magneti Marelli, 159
DiSalvo, Jason, 229
Domenicali, Claudio, 233, 235
Donington, 156, 166
Dovizioso, Andrea, 235, 236, 238–239, 242, 243, 247, 256, 260, 264
Drudi, Aldo, 193, 223
Drusiani, Alfonso, 23
Ducati, Adriano Cavalieri, 6–7
Ducati, Bruno, 6–7
Ducati, Marcello, 6–7
Ducati Elettrotecnica, 19
Ducati Meccanica, 19
Dunne, Carlin, 231, 235, 260
Dunscombe, Alan, 108

E

EFIM Group, 94, 112, 137, 149
Emmett, Sean, 192
Endurance World Championship, 149
Enduro, 128
European Superbike Championship, 163

F

F750 Imola, 109
Fabbro, Gianni, 213
Fabrizio, Michel, 216, 221, 225
Factory Superbike, 195
Falappa, Giancarlo, 165–166, 173
Farinelli, Aldo, 9
Farnè, Alberto, 21
Farnè, Franco, 21, 29, 35, 39, 48, 76, 81, 103, 113, 137, 139, 157
Fattorino Carrier, 75
FB Mondial, 23
Ferrari, Virginio, 149, 151, 179, 182
Ferri, Romolo, 39, 40
FIM Superstock World Cup, 214, 218, 223, 229
Findlay, Jack, 108
Finmeccanica, 137
Fiorio, Giovanni, 11, 13, 14, 16, 18
Fogarty, Carl, 166, 167–168, 173, 177, 182, 185, 188, 190, 193
Fontana, 102
Forcella Italia, 152
Forcinelli, Aldo, 27
Foyt, A. J., Jr., 74
Francini, Vanes, 137
Frasers, 133

G

Gallina, Roberto, 85
Galluzzi, Miguel, 173, 183
Gandossi, Alberto, 21, 29, 31, 34, 35, 39, 40
Garriga, Juan, 146, 151
Ghia, 17
Gibernau, Sete, 208
Gilera, 39
Girino, 11
Girling, 55
Giro d'Italia, 23, 37
Giugiaro, Giorgetto, 122, 123
Giugliano, Davide, 229, 239, 240, 243, 247
Giuliano, Ermanno, 102, 108, 233
Gobmeier, Bernhard, 235
GP10, 225
GP12, 228
GP13, 235
GP15, 242, 243
GP16, 247, 251
GP17, 251, 256
GP18, 256, 260
GP19, 260, 264
GP20, 264, 265
GP21, 269
Gran Fondo, 34

Gran Sport (Marianna), 23, 29
Grand Prix, 35, 39
 125, 41
Grant, Mick, 112
Grau, Benjamin, 112, 121, 143, 146
Graziano, Antonio, 29, 34
Grimeca, 36, 105
GT1000, 215

H

Haga, Noriyuki, 221, 225
Hailwood, Mike, 41, 46, 55–56, 130–132
Hailwood, Stan, 46, 54
Harley-Davidson, 67, 69
Hayden, Nicky, 221–222, 225, 226, 228, 230, 233, 235, 238
Hislop, Steve, 177, 192, 195
Hodgson, Neil, 190, 199, 200
Hypermotard, 213, 218, 224, 230–231, 234–235, 236–237, 241, 247
 796, 225, 227–228
 939, 248, 250
 939SP, 248, 250
 950, 260, 262
 950SP, 262, 274
 1100, 214–215, 224
 1100 EVO, 227–228
 1100 EVO SP, 227–228, 235
 1100S, 214–215, 224
 S, 247
 SP, 225
Hyperstrada, 235, 236–237, 241, 247, 250
 929, 247
 939, 248, 250

I

Iannone, Andrea, 239, 242, 243, 247
Iddon, Christian, 265
Imola, 23, 101, 107, 141
 200, 108, 112
 Replica, 113, 116
Irwin, Glenn, 251
ISDT, 13, 21
Isle of Man, 40, 130–131, 132, 134, 141
Italian F1 Championship, 149
Italian Internationals, 102
Italian Senior Championship, 102

J

Juan, Enrique de, 143
Junior Italian Championship, 29, 35, 39, 137, 139, 141

K

Kavanagh, Ken, 56
Kayaba, 244–245
Kneubuhler, Bruno, 112
Kocinski, John, 179
KTM 1290 Super Duke, 248

L

Laconi, Régis, 203, 205
Laconia, 48
Lafranconi, 122, 126, 135
Laguna Seca, 150, 156, 226, 242
Lambretta, 75
Landi, Silvio, 21
Lanzi, Lorenzo, 209, 213
Lavilla, Gregorio, 205
Lelli, 23
Leoni, Reno, 85
Lewis, Johnny, 250
Limited Edition 1199 Superleggera, 238
Limited Edition Fogarty S4, 192
Limited Edition Laguna Seca, 154, 156
Lockheed, 113
Lorenzo, Jorge, 251, 256
Loria, Aldo, 10
Lucchinelli, Marco, 148, 149, 150, 151, 155, 156, 158

M

M800, 202
M900 I.E., 197
M900S, 188
M1000, 202
M1000S, 202
Malaguti, Giovanni, 21

INDEX | **279**

Malaysian Grand Prix, 206
Mallol, Jose, 137
Mandolini, 34
Mandracci, Guido, 108
Maoggi, Giuliano, 26, 28
Marchesini, 217, 227, 237, 239
Marconi, Guglielmo, 6
Marelli, 204
Marianna, 20, 23
Marinelli, Ernesto, 239
Marini, Luca, 269, 273
Mark 3 Desmo, 90, 92, 93
Marlboro, 48
Martin, Jorge, 269, 273
Martin, Steve, 188
Marvic, 162
Marvic/Akront, 153, 157, 158, 167
Marzocchi, 42, 45, 56, 59, 77, 85, 91, 92, 93, 102, 103, 105, 107, 115, 117, 130, 154, 162, 237
Melandri, Marco, 216, 251, 256
Menchini, Pietro, 137
Mengoli, Gianluigi, 136
Mercado, Leandro, 240
Mertens, Stéphane, 163
MH900e, 190, 192, 195
Michelin, 157
Micromotore class races, 13
Mike Hailwood 900 Evoluzione, 195
Mike Hailwood Replica, 131, 134
Mikuni, 162, 164, 191
Milan Show, 18, 19, 21, 34
Milano-Taranto, 13, 15, 23, 28, 29, 34
Mille, 145
  Mike Hailwood Replica, 147, 149
  S2, 147, 149
Miller, Jack, 256, 260, 264, 269, 273
Miller, Sammy, 39
Milvio, Arnaldo, 94, 96, 98, 102, 112
Minoli, Federico, 182, 205
Misano, 150
Misano Autodromo Santamonica, 158
Modena, 81
Mondial, 39
Monetti, Giorgio, 37
Monster, 173, 190, 191–192, 193, 195, 196–197, 201–202, 204–205, 206–207, 209, 215–216, 218, 224, 225, 228, 230, 234–235, 238, 241, 244–245, 270, 271
  659, 231
  695, 208, 216
  696, 218–219, 220, 224, 228, 230, 238, 241
  795, 233
  796, 230–231, 238, 241
  797, 254
  821, 241–242, 244–245, 259–260
  1100, 221, 224, 228
  1100 EVO, 230–231
  1100 EVO SP, 238
  1100S, 224, 228, 230
  1200, 238, 241–242, 244, 254
  1200R, 247, 248
  1200S, 241–242, 254
  1200S, 244
  S2R, 207, 209–210, 215, 218–219
  S2R1000, 208
  S4, 192
  S4 Fogarty, 193–194
  S4R, 204–205, 207, 209–210, 215, 219, 224
  S4RS, 208, 209–210, 215, 218–219, 224
  Special Edition Tricolore S4RS, 221
  Stripe 1200, 244
Montanari, Alano, 29
Montano, Giuseppe, 15, 19, 23, 29, 37, 67, 94
Montjuïc, 143, 152
Montjuïc 24-hour race, 34, 52, 112, 121, 137
Monza, 50, 151
MotoE™ World Cup, 277
Moto Guzzi, 39, 108
Motogiro d'Italia, 21, 23, 24, 26, 29, 34, 35
MotoGP, 195, 198, 199, 203, 205, 209, 212, 216, 221, 225, 228, 233, 235, 242, 247, 251, 256, 264–265, 269, 273, 277
Mototrans, 107
Multistrada, 201, 203–204, 206, 208–209, 211, 212, 220, 226, 227, 232–233, 236–237, 243

950, 253–254, 263
950S, 253
1100, 215, 224
1100S, 215
1200, 213, 226–227, 230, 237–238, 244, 245
1200 Enduro, 248, 249
1200 Enduro Pro, 253–254
1200S, 226–227, 244, 245
1200S Pikes Peak, 231
1200S Touring, 235
1260, 258
1260 Enduro, 263
1260 S, 258
1260 S D-Air, 258
1260 S Grand Tour, 266–268
1260 Pikes Peak, 258
D-Air, 245, 248
Pikes Peak, 248
S, 248
S Granturismo, 237
S Pikes Peak, 237
S Touring, 237–238
U-Air, 244
V2, 274
V2S, 274
V4, 270
V4 Pikes Peak, 274
V4S, 270–271
V4S Sport, 270–271
MV Agusta, 102, 108

N
Nardi and Danese, 32
Nations Grand Prix, 31, 35, 40, 48
NCR, 113
NCR Classic, 134
Neilson, Cook, 113, 120, 128
Neri, Renzo, 103, 113
Nippon Denso, 128, 129, 134, 136
Norton, 55

O
Öhlins, 161, 163, 169, 191, 193, 196, 206, 207, 209, 210, 213, 215, 217, 224, 229, 234, 236, 237, 239, 241, 244, 248, 250
Oldani, 63, 64, 73, 81, 85
Olivero, Maestro, 10
Oscam, 143, 144, 152, 153, 156

P
Pagani, Alberto, 108
Paioli, 129
Panigale, 233
  899, 241
  899 Supermid, 240
  959, 248, 249
  1199R, 239
  1199RS13, 235–236
  R, 242, 243, 244, 247, 251, 256
  S Senna, 240
  V2, 266, 273
  V2 Bayliss 20th Anniversary, 273
  V4, 257, 266, 267, 269, 276
  V4 25th Anniversary 916, 261–262
  V4R, 260–261, 265, 266, 267, 269
  V4S, 257–258, 266, 269, 276
  V4S Corse, 261
  V4 SP, 269–270
  V4 Speciale, 257
  V4 Superleggera, 265–266
Parlotti, Gilberto, 85
Paso, 150
Pasolini, Renzo, 151
Pedretti, 148
Petrucci, Danilo, 256, 260, 264
Petrucci, Franco, 13
Pian delle Fugazze, 29
Pikes Peak International Hill Climb, 231, 235, 260
Polen, Doug, 163, 165, 168
Preziosi, Filippo, 199, 235
Promotor 996, 179–180
Provini, Tarquinio, 39
Purslow, Fron, 41

R
Rabat, Tito, 260, 264
Radaelli, 57, 118
Recchia, Mario, 13, 15, 21
Redding, Scott, 256, 261, 265, 269
Reyes, Luis, 143, 146

Reynolds, John, 192
Ricard, Paul, 146, 151
Ricardo, 102
Rinaldi, Michael Ruben, 251, 265, 269, 273
Roberts, Brendan, 219
Roche, Raymond, 160, 161, 163, 165
Rolly, 88–90, 94
Rossi, Valentino, 228–229, 230, 233
Rosso, Renzo, 238
Rutter, Tony, 139, 141, 143, 144, 145

S
S4, 202
Saccomandi, Sergio, 21
Sachs, 215, 237, 244–245
Sanremo Ospedaletti, 102
Scamandri, 29
Scarab, 113, 117
Schilling, Phil, 113, 128
Schwantz, Kevin, 240
Scrambler, 241, 245–246
  1100
  1100 Pro, 268
  1100 Special, 259
  1100 Sport, 259
  1100 Sport Pro, 268
  1100 Tribute PRO, 275
  Café Racer, 255, 263
  Desert Sled, 255, 263
  Fasthouse, 273
  Flat Track Pro, 248–250
  Full Throttle, 250, 263–264
  Icon, 263
  Mach 2.0, 259
  Nightshift, 272
  Sixty2, 248–250
  Street Classic, 259
  Urban Motard, 275
Scuderia NCR, 131
Sebring, 76
Seeley, Colin, 101, 102
Senna, Ayrton, 177
Senna I, 179
Senna II, 182–183
Sestini, Marcello, 35
Showa, 164, 169, 187, 188, 191, 240
Silentium, 36, 135
Siluro, 32
Silver Shotgun, 104
Simeon, Xavier, 223
Skyhook Suspension, 237
Slinn, Pat, 131
Smart, Paul, 108, 109
Spaggiari, Bruno, 29, 34, 39, 102, 108, 109, 112
Spairani, Fredmano, 94, 96, 98, 102, 108, 112, 113
Spanish Mototrans, 96
Speedline, 132, 135
Sport
  1000, 215
  1000S, 215–216
  GT1000, 215
Sport Touring, 193, 196–197, 201–202, 204, 206–208, 209, 215
SportClassic, 210–211, 215, 218, 224, 225
  1000S, 224
  GT 1000 Touring, 224
  GT1000, 211, 224, 228
  Paul Smart 1000 Limited Edition, 210–211, 215
  Sport 1000, 211, 215
Sprayson, Ken, 55
Sprayson, Reynolds, 56
SSI carburetor, 52, 77, 78
ST2, 182, 183, 188, 189, 191–192
ST3, 204, 209–210, 215
ST3ABS, 215
ST3sABS, 209–210
ST4, 188, 189, 191–192, 193, 202, 204
ST4S, 192, 193, 194, 204
ST4sABS, 202, 209
Stoner, Casey, 212, 213, 216, 217–218, 221, 225, 247
Streetfighter, 213, 221, 224, 225, 228, 230, 234–235, 238, 241, 247
  848, 233, 235, 241, 247
  S, 224
  V2, 275
  V4, 260, 266, 267
  V-4 prototype, 260

V4S, 266, 267
V4SP, 275
Superbike, 225
Supermono, 167, 169–169, 173, 177
Superquadro, 233–234
Supersport, 190, 193, 196–197, 201–202, 204, 206–208, 209, 210, 252–253, 271
Surtees, John, 55, 56
Swedish Grand Prix, 31

T
Taglioni, Fabio, 21, 22, 23, 28, 29, 30, 35, 54, 67, 69, 76, 79, 90, 96, 98, 103, 108, 109, 113, 116, 122, 124, 127, 136, 137, 148, 208, 233, 251
Tamarozzi, Ugo, 13
Tamburini, Massimo, 151, 159, 173
Tardozzi, Davide, 163
Tartarini, Leopoldo, 27, 34, 37, 92, 113, 115, 127, 128, 129
Taveri, Luigi, 39
Tejedo, Alejandro, 137
Terblanche, Pierre, 166, 170, 187, 190, 192, 195, 199, 201, 210, 213
Termignoni, 193, 236
Testastretta, 190, 192, 193, 195, 196, 199, 200, 204, 230, 235, 236, 241, 244
  DVT, 248
  Evoluzione, 213, 224, 226, 229
Texas Pacific Group, 173, 179
Toseland, James, 203, 205
Tracy, Greg, 231
Trellis, 241

U
Ubbiali, Carlo, 31, 39
Ulster, 46, 139, 141

V
Veglia, 64, 78, 152
Verlicchi, 131, 146, 152, 159
Vespa, 75
Vighi, 23
Vignelli, Massimo, 185
Vila Real, 141
Villa, Francesco, 23, 39, 40, 48, 54
Villa, Walter, 146
VM Group, 137, 143
Volkswagen Group, 233
VR46 Racing Team, 273

W
Weber, 13, 151, 159, 162, 164, 177
Weber Marelli, 190
World Championship, 39, 46
World Superbike Championship, 156, 160, 161, 163, 165, 167, 173–174, 176–177, 179–180, 182, 185, 186, 188, 190, 192, 195, 199, 200, 201, 203, 205, 206, 208–209, 215, 216, 221, 223, 225, 228–229, 235, 239, 242, 247, 251, 256, 260, 269, 273
World TT Formula 1 Championship, 131, 150, 158
World TT Formula 2 Championship, 139, 141, 143, 144, 145
Wynne, Steve, 131

X
Xaus, Ruben, 192, 200, 214
XDiavel, 247–248
XDiavel Black Star, 272
XDiavel Dark, 272
XDiavel S, 247–248

Z
Zaiubouri, Franco, 125
Zarco, Johann, 264, 269, 273
Zitelli, Glauco, 13